Organic Vegetable Production

A COMPLETE GUIDE

Organic Vegetable Production

A COMPLETE GUIDE

Edited by Gareth Davies and Margi Lennartsson

In association with the Henry Doubleday Research Association

THE CROWOOD PRESS

First published in 2005 by
The Crowood Press Ltd
Ramsbury, Marlborough
Wiltshire SN8 2HR

www.crowood.com

British Library Cataloguing-in-Publication Data
A catalogue record for this book is available from the British Library.

ISBN 1 86126 788 6

ACKNOWLEDGEMENTS

The editors wish to thank the numerous people involved in bringing this book to publication and we are grateful to all of the authors for their numerous contributions. In writing and editing the book we have drawn on our collective experience in organic research over the previous decade or so and as such we have relied on the knowledge, experience and advice of many people in the 'organic movement' and beyond. To them all, too numerous to mention, we would say thank-you. A special mention must go to DEFRA whose funding for organic research has underpinned much of HDRA's research work. We would also especially like to thank all the organic farmers and growers with whom HDRA has worked, and who have willingly contributed their stories, knowledge, experience and time to our research programmes. They have made a real contribution to the information contained in this book, and, without them, the book would not have been possible.

Typeset by SR Nova Pvt Ltd., Bangalore, India

Printed and bound in Great Britain by The Cromwell Press, Trowbridge

Contents

Foreword

It is a matter of history that the production of organic vegetables did much to lay the foundations for what is now a significant market for all types of organic produce. The appearance of organic carrots on the shelves of a major supermarket in 1985 was for many consumers their first encounter with organic produce. Over the intervening years organic vegetables (and to some extent fruit) have provided an entry point to the organic market for many thousands of consumers, and for some they represent the only organic produce they recognize and buy. Organic vegetable sales do not dominate the market as they did in those early days, but they are still extremely important and significant.

Contrary to some popular perception in past years organic vegetable production is not about 'muck and magic', though of course muck can sometimes play a significant role if properly sourced and composted. It is crucially dependent on hard work, attention to detail and an ability to face the slings and arrows of the marketplace with determination and persistence. It is also crucially important to have a good understanding of basic horticultural processes and techniques. In adopting an organic approach to food production, growers also need to have a good understanding of the relationships in the system and how to balance them for maximum effect. These might include relationships between predators and pests, soil structure and biological activity, weeds and bio-diversity, among many others.

From the early days to the present time information was and is a key commodity for new entrants to the industry, both converting conventional growers and established organic producers. The sector is more complex than any other given the range of crops grown, the varying degrees of intensity of production and the wide range of soils in which crops are grown. This last topic is set to become even more important given the increasing importance of local production for local markets. Historically, information on organic vegetable growing has been largely based on grower experience. This was a contributory factor in the establishment of the Organic Growers Association in 1979, a role that has been superseded by its absorption into Soil Association Producer Services in 1991. The need for information and advice was also instrumental in the setting up of the

Organic Advisory Service at Elm Farm Research Centre. OAS horticultural advisors are still actively providing a service to the industry today.

Grower experience has also played a very important role in what has been a relatively small number of books published so far aimed at the commercial organic grower. These books have played and continue to play an important role in the provision of information. The book that you are about to read represents a step forward in that it combines grower experiences with a wealth of information gleaned from the many research and development projects undertaken by the HDRA over the last 10–15 years. These projects have covered a range of topics that includes variety trials, green manures, soil fertility, nutrient management, field-scale conversion, weed management, pest and disease control, and many more. Several of these projects have involved growers across the country on holdings of varying size and supplying all levels of the market. This book represents a unique bringing together of information by a committed team of researchers and advisors. You may sit down and read it from cover to cover or, more likely, you will use it as a work of reference that informs all aspects of the process of successfully producing organic vegetables on a commercial holding. However you approach it, you will find it very useful.

Roger Hitchings
Head of Advisory Services, Elm Farm Research Centre

ACRONYMS AND ABBREVIATIONS

ACOS – Advisory Committee on Organic Standards

ADAS – (formerly) Agricultural Development and Advisory Service

COSI – Centre for Organic Seed Information

CSL – Central Science Laboratories

DEFRA – Department of Environment, Food and Rural Affairs

EFRC – Elm Farm Research Centre

ERDP – England Rural Development Programme

FYM – farmyard manure

HDC – Horticultural Development Council

HDRA – Henry Doubleday Research Association – the Organic Organization

IFOAM – International Federation of Organic Agriculture Movements

NIAB – (formerly) National Institute of Agricultural Botany

OAS – Organic Advisory Service

OCIS – Organic Conversion Information Service

OCW – Organic Centre Wales

OELS – Organic Entry Level Scheme

RES – Rural Enterprise Scheme

SA – Soil Association

UKAS – United Kingdom Accreditation Service

UKROFS – United Kingdom Register of Organic Food Standards

Warwick HRI – Horticulture Research International, University of Warwick

Chapter 1

Organic Principles for Growing Vegetables

Organic agriculture is a system of food production that acknowledges the importance of biodiversity, soil biological activity and biological cycles. It aims to enhance and develop these as an integral part of the food-production system. Organic systems also recognize that human health is linked to the type and quality of the food that people eat and that this, in turn, can be directly linked to the health of plants and animals that provide this food and, ultimately, to the well-being of the soil on which all agricultural production depends. Organic agriculture also regards animal welfare and human social well-being as integral to agricultural production. The many, diverse, organic farming systems share similar practices or characteristics that stem from this underlying viewpoint. More recently, these perspectives are increasingly being set out as 'organic principles' or even legal requirements. The principles are briefly described below in order to provide the framework for understanding many of the practices for organic vegetable-production described in the subsequent chapters.

PRINCIPLES FOR ORGANIC AGRICULTURE

The general principles and ideas underlying organic agriculture are now, at least superficially, well known. In restating these principles, we have largely followed those developed by the International Federation of Organic Agriculture Movements (IFOAM). These are not intended to be a proscriptive recipe for organic farming but a flexible set of principles that can be adapted to local circumstances. IFOAM does not prioritize any of the principles and recognizes that they are all important. According to IFOAM, the principles are that organic agriculture should:

- produce sufficient quantities of high-quality food, fibre and other products

- work compatibly with natural cycles and living systems through the soil, plants and animals in the entire production system
- recognize the wider social and ecological impact of and within the organic production and processing system
- maintain and increase long-term fertility and biological activity of soils using locally adapted cultural, biological and mechanical methods – as opposed to relying on inputs
- maintain and encourage agricultural and natural biodiversity of the farm and its surroundings through the use of sustainable production systems and the protection of plant and wildlife habitats
- maintain and conserve genetic diversity through attention to on-farm management of genetic resources
- promote the responsible use and conservation of water and all life therein
- use, as far as possible, renewable resources in production and processing systems and avoid pollution and waste
- foster local and regional production and distribution
- create a harmonious balance between crop production and animal husbandry
- provide living conditions that allow animals to express the basic aspects of their innate behaviour
- utilize biodegradable, recyclable and recycled packaging materials
- provide everyone involved in organic farming and processing with a quality of life that satisfies their basic needs, within a safe, secure and healthy working environment
- support the establishment of an entire production, processing and production chain that is both socially just and ecologically responsible
- recognize the importance of, and protect and learn from, indigenous knowledge and traditional farming systems.

More recently IFOAM has attempted to summarize these principals as:

- *the principle of health*: organic agriculture should sustain and enhance the health of soil, plant, animal and human as one and indivisible
- *the ecological principle*: organic agriculture should be based on living ecological systems and cycles, work with them, and emulate and help sustain them
- *the principle of fairness*: organic agriculture should be built upon relationships that ensure fairness with regard to the common environment and life opportunities
- *the principle of care*: agriculture should be managed in a precautionary and responsible manner to protect the health and well-being of current and future generations and the environment.

The principles flow from an underlying world-view of what agriculture can and should set out to achieve. In a sense, these could be said to be 'ideals'. It is from these ideals that sets of organic standards, against which individual farm systems can be compared, have been drawn up, which has led to the concept of certified organic farming systems.

CERTIFICATION OF ORGANIC FARMING SYSTEMS

Certification of organic farming systems has arisen as a means of evaluating working practices of farmers and processors, to assess whether they are applying and adhering to organic standards – hence following organic principles – in their production systems. Generally, the aim is to certify the production system rather than the end product, although, in the eyes of the consumer at the end of a long food-chain, this is often overlooked. Certification is, therefore, a means of ensuring the integrity of organic products and has subsequently become subject to legal codification (*see* below). Consumers are assured that certified organic produce has been produced, stored, processed, handled and marketed in accordance with organic standards, or technical specifications, which have been certified as organic by a recognized body.

In the UK and across Europe, there are numerous organic certification bodies. These interpret and codify organic principles in 'organic standards'. Obviously there will be slight differences, even within countries, in the detail of the standards as interpreted by each certification body. As a minimum, they do have to conform to EU and national laws (*see* below) for organic production and processing. In the UK, organic certification bodies conform to the Advisory Committee on Organic Standards (ACOS) standards, which are based on the EU law (regulation 2092/91 and amendments), and, in turn, they are approved by ACOS as certification bodies. ACOS is a non-executive non-departmental public body within the Department of Environment, Food and Rural Affairs (DEFRA), which advises ministers on matters related to organic standards. It was preceded by the United Kingdom Register of Organic Food Standards (UKROFS).

Legal Requirements

In 1993, EC Council Regulation 2092/91 became effective across the EU and, since that time, organic food production has been legally regulated in the EU. Regulation 2092/91 sets out the inputs and practices that may be used in organic farming and growing, and defines the inspection system that must be put in place to ensure this. This regulation also applies to processing, processing aids and ingredients in organic foods. Therefore, all foods sold as organic must originate from growers, processors and importers who are registered with an approved certification body and subject to regular inspection. Regulation 2092/91 has formed the basis for UK organic standards, initially in the form of UKROFS standards. These have been replaced by the Compendium of UK Organic Standards.

Practical Requirements

All organic producers have to be registered with a certifying body, and be 'certified' in order to sell their produce as organic. The government, through the government Department for Environment, Food and Rural Affairs (DEFRA) and the United Kingdom Accreditation Service (UKAS), oversees an increasing number of these inspection or certification bodies (thirteen as of December 2004). Each inspection body has its own set of organic standards that have to conform to the EU law, which, in turn, sets the legal standards. Some certification bodies may operate standards that are 'stricter' than this official or legal minimum, but these differences are generally more pronounced in the livestock sections of the standards than in the horticulture sector.

The process of certification involves some form of farm visit by inspectors of the certifying body, in which the production system is visually inspected, together with a review of the business practice and accounts, normally on an annual basis. It is the role of the inspection body to ensure that growers follow the set standards, so that consumers can have confidence in the authenticity of organic food. Testing of the soil or produce is not generally part of the process, though the certifying body can employ it if a problem is suspected. Detailed and accurate record-keeping is absolutely fundamental to the operation and inspection of organic systems.

Once certified, growers have the right to use the certifying agency's label. This enables the consumer to find out under what conditions they can expect their food to have been produced by consulting the certifier's standards.

CHARACTERISTICS OF ORGANIC FARMING SYSTEMS

Whilst there might seem to be a growing body of rules and regulations surrounding organic agriculture, the principles and the practice work to ensure that certified organic farming systems share certain characteristics. We have described these briefly below, so as not to loose sight of what features of organic systems are important for all stakeholders, be they farmers, processors or consumers.

Organic farming systems are generally working towards increasing biodiversity. Biodiversity aids production through many biological services, such as pollination, natural pest-control and erosion control. Organic farmers are generally interested in building biodiversity into their farming systems and consumers desire the type of food and the landscapes this creates. Elements in farming systems that enhance biodiversity include rotations, intercropping and sensitive habitat-management and many of these are described in this book.

Organic farming systems are working towards integrating economic, environmental and social sustainability into their farm systems. Sustainability recognizes that there are limits to growth in a finite ecosystem and that the goal of food-production systems should be to develop ecological and social resources that maintain or enhance our future ability to produce food, either for future generations or ourselves. This naturally means that organic farmers should be working towards reducing external inputs, be it fuel or feedstuff, into their farm systems and looking to produce all resources within the farm boundaries or near locality.

Organic farming systems generally concentrate on maintaining a healthy and fertile soil, capable of supplying adequate and natural nutrition to crop plants. This has many beneficial effects throughout the whole farm system and many of the techniques for maintaining soil fertility are alluded to in later chapters. Many organic proponents, including the founders of the modern organic movement like Eve Balfour, go further and make a link between a biologically healthy soil, the production of wholesome food and, ultimately, human health.

Organic farming systems rely on natural nutrient-cycles to provide crop nutrition and more natural, preventative methods of pest, disease and weed management. In both cases organic farmers rely on natural ecological processes to provide services that are normally bought in as 'inputs' in conventional farming systems. In practice, organic farming strictly restricts the use of artificial chemical fertilizers and synthetic pesticides. Animals are reared without the routine use of drugs, antibiotics and de-wormers and should be allowed to display more natural behaviours.

Biodiverse, sustainable farming systems naturally lead to diverse farm enterprises. Organic farmers often integrate various farm enterprises and, typically, organic farms run more complex and labour-intensive farm operations. Many are also involved in more direct marketing of produce to local consumers and even, in some cases, linking consumers with crop production. Such diverse enterprises are also capable of meeting many of the social principles of organic farming and wider rural development initiatives.

An organic farm is part of the ecological and social landscape

Above all, most practitioners of organic agriculture would describe their practices as holistic. That is, integrating the various practices and management methods so that the farm operation becomes a system of interlocking and interactive parts that work together.

AIM OF BOOK

The subsequent chapters of this book aim to give an overview of current, appropriate, practice in organic vegetable and potato production systems. It will cover the general principles of organic-vegetable production, starting with an overview of organic vegetable production systems in the UK and the conversion of conventional to organic production practices, in the latter case indicating potential problems that might arise. It will then develop the key production themes, including crops, fertility building, weed-control and pest and disease management. These themes have then been brought together under a series of chapters that consider the practical farm-management aspects of production, including planning for successful rotations, marketing and practical vegetable production from the point of view of specific crops, highlighting the specific characteristics or requirements of various crop-types. The book will also briefly consider post-harvest aspects, such as storage and protected cropping – increasingly important as techniques for extending the marketing season. The penultimate chapter looks at the importance of acquiring knowledge, together with potential sources of information and advice, before a final chapter that develops a brief overview of likely future trends. In Further Reading and Useful Addresses, sources of information are also provided.

Chapter 2

Organic Vegetable Production Systems

Organic vegetables are grown in a wide range of organic farm systems. Whilst it is generally recognized that each organic farm is, in many senses, unique, it is also possible to classify various broad types of farming systems in which organic vegetables are likely to be produced. This can give a useful indication of the opportunities that exist for growing organic vegetables. It is also helpful because the type of system will, to some extent, dictate the management options open in any particular situation.

The previous chapter has described the principles that underlie the design of organic farming systems. As a consequence of these principles (and the standards based on them), crop rotation is usually seen as the heart of any organic farming system. Here we categorize the organic farming systems in which organic vegetables are generally grown into four broad categories, based on the type of rotation, and briefly examine each. These are intensive vegetable rotations, arable and vegetable rotations, mixed-farming rotations and novel or non-traditional systems. An outline scheme or rotation for each system is given and then the management, agronomic and marketing factors of each are briefly touched upon. Many of these aspects are developed in more detail in forthcoming chapters.

All organic vegetable systems need to be legally certified and, in order to achieve this status, all farms will need to undergo a period of conversion to organic production. This conversion phase is a period that can involve a considerable amount of investment, both in equipment and learning. The various issues that are likely to arise during conversion are discussed in the latter part of the chapter.

VEGETABLE PRODUCTION SYSTEMS

Intensive Vegetable Systems

Only vegetables (in a broad sense) are grown in intensive vegetable systems and no arable or livestock are included in the rotation. A typical

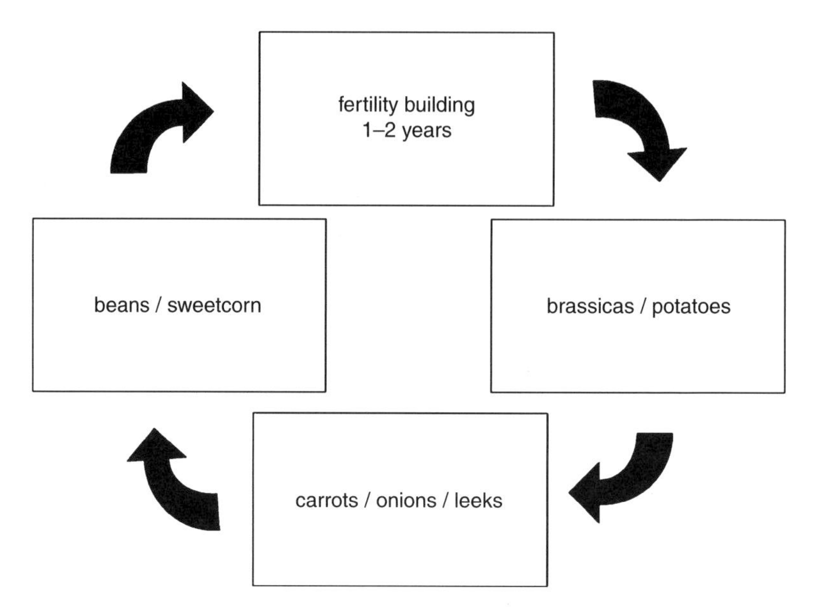

Figure 2.1 A typical intensive vegetable rotation

system (*see* Fig. 2.1) may be based on a four- or five-year rotation with a one- or two-year fertility-building phase.

No vegetables or cash crop are grown in the fertility-building phase, only species that fix nitrogen, such as clover or vetch, in order to replenish soil fertility. Nitrogen-fixing plants are commonly grown in combination with nitrogen-lifting plants in this phase. The latter are deep-rooting plants, often grass species, which take up nitrogen and prevent it from being leached out of the rooting zone and also add to the soil organic-matter. The most common combinations are grass and clover, and grazing rye and vetch. In the absence of grazing, such crops will need to be mown frequently during the summer and the cuttings left on the field.

Vegetables are grown in the remaining part of the rotation. Crops with high demand for nitrogen, such as brassicas or potatoes, are normally grown in the first year after fertility building. In the second year, crops with a lower demand for nitrogen, such as carrots, onions, leeks or celery, are grown. A third year of vegetable cropping may include beans or sweetcorn.

Such rotations are typical of high-value land suited to vegetable production, such as the fen-type soils found in Lincolnshire or Lancashire. Paradoxically, such farming systems are often suited to either very small or very large producers.

Small market-gardens or smallholdings are most likely to adopt the intensive vegetable system and sell produce directly through farmers' markets or to local wholesalers. Such a marketing system has the advantage of having a short supply chain, so that the grower retains a higher proportion of the retail price. Small market-gardens often grow a wide diversity of crops in small areas. Although this is labour intensive, diversity has the advantage of preventing the spread of pests and diseases and reduces the impact of crop failure. The disadvantage of such systems is that local markets can only accept limited volumes of produce and so it is unsuitable for larger farms.

Large-scale field vegetable production systems usually sell produce to packers who supply supermarkets, as these are the only outlets that will accept large volumes of produce. These systems will have all the advantages of economies of scale and will often have the capital to invest in considerable mechanization, thereby increasing the efficiency of production. The disadvantages of selling produce to packers are that the supply chain is longer and the grower usually receives a lower price. Quality standards are often very stringent, so the risk of a crop being rejected is high. This is compounded by the fact that supermarkets often over-programme to ensure that sufficient produce is available. The resulting glut in the market puts packers in a position to reject a crop on seemingly superficial grounds.

Arable Rotations Including Vegetables

Arable rotations incorporating vegetable production can span a range of systems, from those that are predominantly arable to those that are predominantly vegetable production systems. At the one extreme the rotation may be similar to that of the intensive vegetable production system (described above), except that a cereal crop, often under-sown with the fertility building crop, will generally be incorporated at the end of the vegetable production phase. At the other extreme, an organic vegetable crop, often potatoes or brassicas, are grown in an arable rotation of cereal, beans and fertility-building leys (*see* Fig. 2.2).

At these extremes the cereal or vegetable crop respectively acts as a break crop in the rotation. Such rotations may be on lower-grade land not so suited for intensive growing of high-value horticultural crops. It may include farms on heavier soils that are not so suitable for growing root crops. In a predominantly arable system, irrigation facilities may not be available, which would preclude the growing of certain types of vegetable crops. Due to the nature of growing combinable crops, this type of system is obviously not suitable for market-garden-scale enterprises but will be found on medium- or large-scale holdings. Despite a drop in the price of

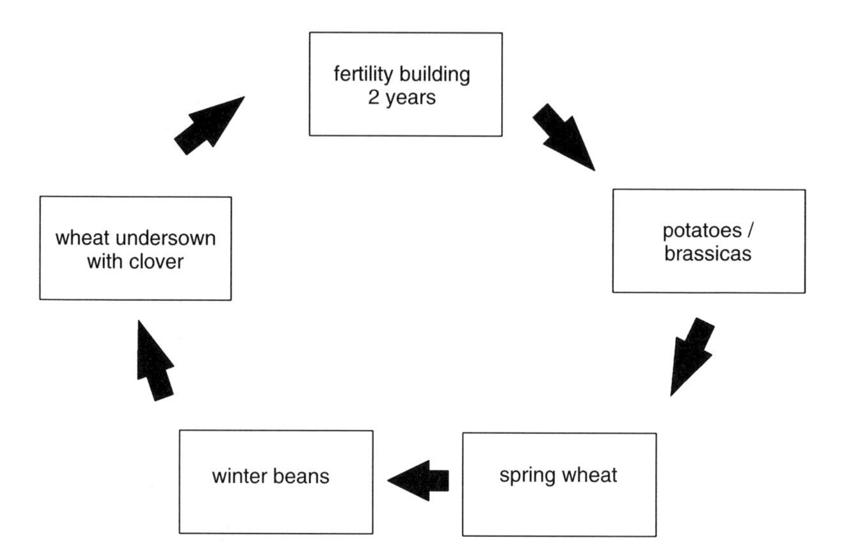

Figure 2.2 An arable rotation that includes vegetables

cereals in 2002, the market is now growing and it is likely to increase further after 2005, when the derogation allowing a proportion of non-organic feed in organic systems was lifted.

Mixed Rotations

Mixed rotations are essentially arable rotations that include vegetables and livestock (*see* Fig. 2.3). In this case the fertility-building period is usually used for silage production and for grazing. They can also include dairy systems with at least semi-permanent pasture. These farming systems also have readily available supplies of animal manure, slurry and bedding, which is a valuable resource for soil fertility management.

Traditionally, livestock systems will be on poorer quality land found in the wetter regions of the country, such as the North West and Wales. Such systems include a range of farm sizes, although a minimum area will be required to allow rotational grazing and include a minimum economic herd size. Such a system must include considerations such as the distance to the nearest organic abattoir.

Novel and Traditional Vegetable-Production Systems

With the development of the environmental and sustainability movements in recent decades, coupled to development of movements that promote

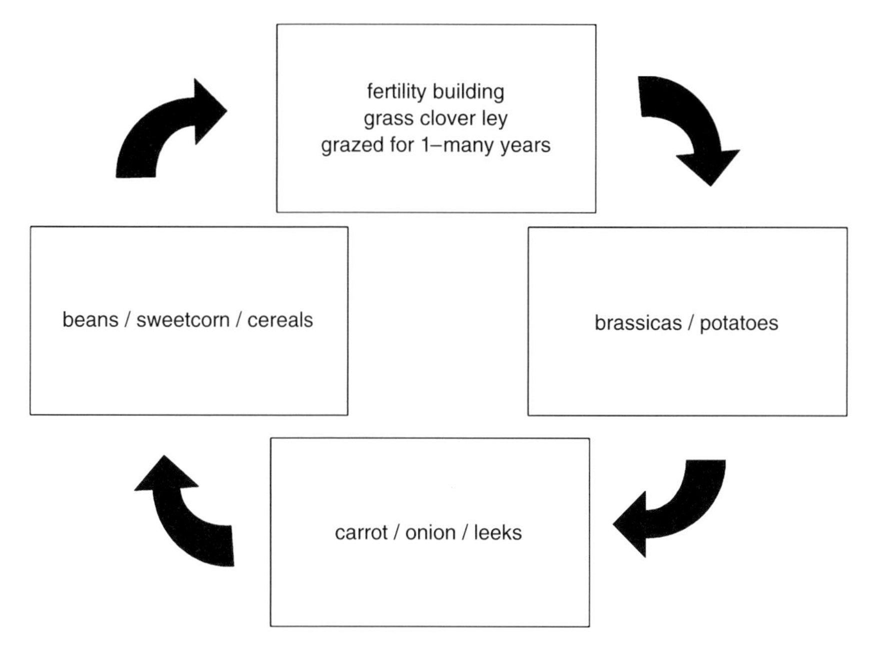

Figure 2.3 A possible mixed-livestock system that includes vegetables

alternative lifestyles, a range of novel 'organic'-farming systems have become increasingly recognized. Many of these systems arise either from philosophical reflection or through trying to put ecological theory into practice, or a combination of the two. Most of these systems represent a development of organic principles as applied to both agriculture and social systems, often called agroecology. Three of these are mentioned below. Traditional vegetable production systems still exist over large parts

Parsnips and pigs, a mixed-farm landscape

of the world and their relationship to organic production is touched upon below.

Biodynamic Agriculture

Based on the teachings of the philosopher Rudolf Steiner, biodynamics is a method of agriculture that seeks to actively work with the health-giving forces of nature. It pre-dates, but has been somewhat overtaken by, organic agriculture. Steiner elaborated a philosophical approach to life (called anthroposophy) that emphasized many of the forces within living nature; identifying many of these factors and describing specific practices and preparations that enable the farmer or gardener to work in concert with these forces. Biodynamic farmers need to learn about the underlying philosophy and apply it to their farming practices. The biodynamic agronomic system is underpinned by a more rigorous and spiritual set of beliefs, which make it less susceptible to the diluting influences that have taken a toll in some areas of organic production, and will therefore be of great interest to the more purist elements of the organic movement.

However, from an agricultural point of view, biodynamics is often seen as 'organics plus', as biodynamic farmers follow all the practices of general organic growing, as well as some more esoteric ones of their own. In any case, vegetables would normally only comprise a small part of a biodynamic farm system. Biodynamic farmers take a more spiritual attitude towards nature and try to capture energy forces in various preparations that they produce and which are sprayed in extremely diluted forms (almost homeopathically) onto the land, crops or compost heaps. A great deal of attention is also paid to the energy of the wider universe and detailed calendars are published annually that map out the best times of day (or night) for performing plant-care activities, especially planting and harvesting, in accordance with the phases of the moon. A significant practical difference between organic and biodynamic agriculture is the use of compost. Biodynamic farmers compost all waste-materials with the relevant preparations to activate and enhance the process before adding them to the soil. The biodynamic standards also stipulate that open-pollinated varieties should be used, preferably those propagated in biodynamic systems. This has led to biodynamic farmers becoming involved in plant breeding programmes on their own farms.

Agroforestry

Agroforestry is the growing of arable and/or horticultural crops together with trees or shrubs on the same piece of land – they often include livestock as well. Agroforestry systems differ from traditional forestry and agriculture by concentrating on the interactions between various system components, therefore taking an organic or holistic approach. Such systems can be designed to provide tree and other crop products at the same time as protecting, conserving, diversifying and sustaining economic, environmental and social resources. Agroforestry, therefore, involves combining tree planting with other enterprises, such as grazing animals or cropping, and should produce a range of products, like firewood, biomass, mulch, fodder and other traditional forestry products (such as wood and nuts). At the same time, the trees can also shelter livestock from wind or sun, provide wildlife habitat, control soil erosion and even, in some cases, fix nitrogen to improve soil fertility.

Vegetable production would normally be only a small part of an agroforestry system, which can include silvopasture (mixing trees with pasture and/or forage), silvoarable (mixing trees and arable or horticultural crops), forest farming (cultivating high-value products in forested areas) and forest gardening (producing many products in a complex forest-environment). Techniques include alleycropping (growing crops in alleys between trees), intercropping and using trees as design elements, such as windbreaks, shelterbelts, woodlots and riparian buffer strips. Contour plantings for erosion control and fertility plantings of nitrogen-fixing trees are also possible options.

Permaculture

Permaculture (a contraction of the words 'permanent agriculture') is a system for the designed integration of landscape and people in order to provide food, energy, shelter and other material and non-material needs in a sustainable way. The ethical basis of permaculture rests upon a care of the earth and developing a system in which all life can thrive. As a technique, permaculture derives inspiration and principles from the study of natural systems; it therefore represents a conscious attempt to design and maintain agriculturally productive ecosystems that have the diversity, stability and resilience of natural ecosystems. It often takes the form of an integrated, evolving system of perennial or self-perpetuating plant and animal species, together with human settlements or communities, organized with the goal of producing an efficient, low-maintenance farming system. This integrates plants, animals, people and structures on all scales from home gardens through to large farms and watersheds. Vegetable production would normally only be a small, but vital, part of such a system. It is represented by a worldwide movement of designers, teachers and grassroots activists working to establish communities that restore or enhance natural ecosystems.

There is great emphasis on practices such as rainwater harvesting, reusing greywater, minimizing the need for irrigation, soil management, maximizing the use of perennial plants and intercropping (*see* agroforestry above), recycling all wastes and producing as many inputs (including sustainable energy, biofuels and fencing and construction materials, as well as manures and composts) as possible on site. The theory is applicable to urban as well as rural settings and much is made of urban and suburban food-production possibilities, although obviously there are constraints on how comprehensively the full philosophy can be adopted in those settings. Many of the agricultural techniques are standard practice in organic crop-production, such as the use of green manures and mulching, promoting biodiversity, companion planting, attracting predators, careful rotation of crops and extensive animal foraging. However, from an agricultural point of view, one major criticism is the suitability of agroforestry or perennial systems in a temperate climate. In warmer climates, such as that of Australia, where the system was developed, not only is the season longer but growth rates are also faster and light-intensity levels higher, making the techniques promoted by the permaculture system much more productive. Many of these issues are, however, being addressed in the UK by a growing band of enthusiastic permaculture practitioners.

Traditional Vegetable-Production Systems

In addition to these novel systems, traditional agricultural systems, including large land-areas in the developing world, as well as smaller allotment- or garden-areas in the developed world, have been increasingly recognized as well adapted for low input and sustainable agricultural systems in their own right. Many of these systems include an element of intensive vegetable production. Whilst many of these are not certified as organic in any sense, it is being increasingly recognized that they could benefit from application of organic principles, especially in the area of soil fertility and pest management. Organic methods are becoming a popular starting point for helping to develop many of these systems to meet both crop production and socio-economic goals. For instance, allotment projects have figured prominently in urban regeneration and social-inclusion projects in parts of the UK.

ADVANTAGES AND DISADVANTAGES OF THE VARIOUS SYSTEMS

Management Factors

It is possible to grow organic vegetables in all the system types and each has its benefits and pitfalls in terms of management. Rotations that include livestock have the big advantage that fertility is sourced from within the farm. Grazing of livestock can also be used to maintain grass/clover leys, eliminating the need for mowing. The arable part of the rotation can also be used as a source of feed. The length of time spent in the fertility-building phase is dependent on the area of vegetables being grown in relation to the whole-farm size. In many cases where a smaller area of vegetables is grown on a larger mixed farm, the best fields can be selected for vegetable production. This system also permits a good deal of flexibility. An intensive vegetable production system or an arable system with only one year's fertility building does not allow so much flexibility, as the majority of the system is in production at any one time. Close attention must also be paid to the management of the fertility-building crops, as a poor ley can have a detrimental effect on subsequent crops for a number of years.

Agronomic Factors

The exact nature of the rotations in organic farming systems will depend on a combination of many factors, including vegetable crops grown, the markets and the location. Rotations are generally unique to each farm and will frequently be changed to fit the circumstances. For example, growers in the south-west who specialize in the niche market for winter cauliflower and early potatoes may practise the rotation shown in Fig. 2.4. This rotation has arisen to fit the needs of the specialist niche-market crops. The winter cauliflower can be planted after the early potatoes have been harvested, whilst a spring cereal can be planted immediately after the harvesting of the winter cauliflower.

Stockless systems are heavily dependent on fertility-building crops to restore the nutrient status of the soil. In many cases, additional nutrients will have to be supplied through the addition of farmyard manure (FYM) that will have to be sourced externally. It is essential that a local source of FYM is readily available on such farms, or an alternative such as green-waste compost. FYM from conventional farms can be used, provided it is declared GM-free and has been composted for at least three months or stacked for six.

Each type of system also brings advantages and disadvantages with respect to pests, diseases and weeds. Long-term grass/clover leys may lead to a build up of wireworms that can cause problems in subsequent potato crops. They are also likely to build up a population of slugs on heavier clay soils. In terms of weeds, all types of rotations have the benefit of breaking cycles of annual weeds, as different planting arrangements and different cultivations for each crop favour different weed types.

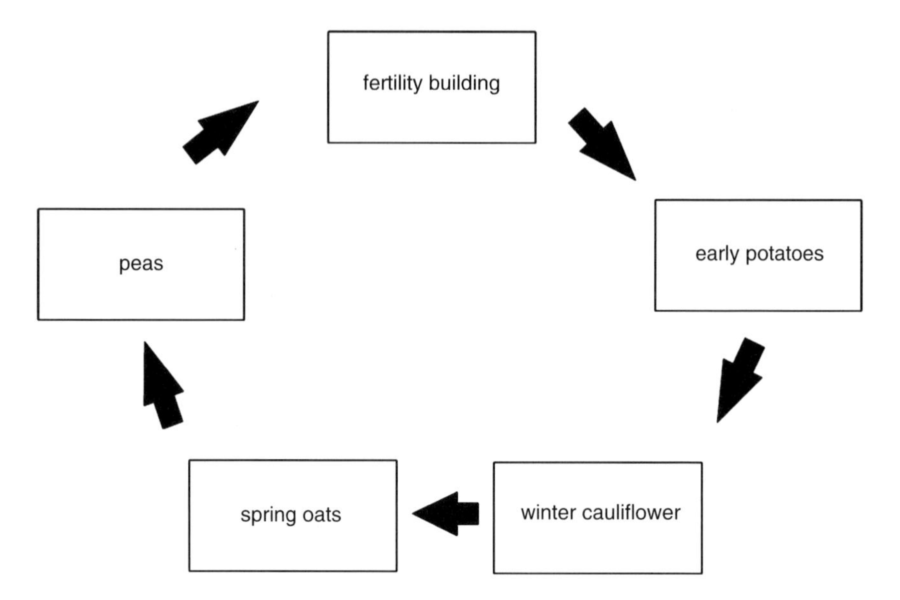

Figure 2.4 An example of a rotation growing early potatoes/winter cauliflower, a speciality or niche market

However, systems with extended periods of grassland may lead to a build up of docks. Animals grazing can have a beneficial effect on controlling weeds and some may be very suited for controlling particular weeds, for example, pigs clearing fields of volunteer potatoes.

Economic Factors

The economics of organic farming should be considered in terms of the whole system and not just one particular crop. In stockless systems, the length of the fertility-building phase determines the proportion of land that is out of production at any one time. It is also important to remember that considerable resources may be required for managing the fertility-building crop. For example, if the ground is dry at sowing, then irrigation may be necessary for proper establishment, as a poor stand in a fertility-building crop may result in nutrient deficiency in subsequent crops. A shorter fertility-building phase in one year means that a greater proportion of the system is devoted to income-generating crops, but may not always be adequate for restoring fertility within the system. In theory, grazing livestock should make economic sense, as the fertility-building phase is being put to economic use and the livestock is generating another source of fertility as FYM. However, in the current economic climate, many organic-livestock enterprises, particularly pigs and dairy, are running at a loss and farmers have found that it is no longer economic to maintain them in the system. Livestock also require a considerable amount of management time and this can have a detrimental knock-on effect on horticultural crops, which can also require considerable management input.

CONVERSION TO ORGANIC VEGETABLE PRODUCTION

The term organic has been defined in European legislation since 1991 and all foods sold as organic have to be certified by a recognized inspection body, as described in Chapter 1. The purpose of the conversion phase is to provide a period of time to establish organic management practices, to build soil fertility and to develop a viable and sustainable agro-ecosystem. Conversion periods are generally two years for grassland and annual-cropping land, and three years for perennial crops in the ground. The certifying bodies have the discretion to shorten the conversion period by up to four months, but this will only be agreed if accurate field records are available and that inspection confirms that no prohibited inputs have been used in the run-up to the start of the conversion process.

Deciding to Convert

Factors to Consider

The decision to convert is often influenced by environmental and/or philosophical motives, but economic factors, related to confidence in the organic market and the state of conventional agriculture, are also important. Consequently, a wide range of factors need to be taken into account before deciding to convert a particular farm or farm system and the ease of conversion will depend on specific farm situations. Here we can only summarize some of the more basic questions that should be asked and these include (roughly in order of importance):

What experience does the grower, or any staff, have of vegetable growing or organic systems? It is a harsh reality that growing organic vegetables is not an easy option and those with prior experience of growing vegetables commercially have tended to adapt more quickly to organic production methods. A considerable amount of effort can be invested in changing management methods during the conversion period and staff may need to be well motivated to see such changes through.

Is the business in good heart economically? Investment is often required, particularly for weed-control equipment, and more will be required if vegetables have not been previously grown or if changes to the marketing are undertaken, for example, packing facilities, vans for distribution. It is important to have a marketing strategy for your organic vegetables. Although government support is often available for conversion, for instance through the Organic Entry Level Scheme (OELS) in the UK, income will be foregone during the conversion period. This is especially true if the land has been under exploitative cropping (such as cereals) and has to begin with a two year fertility-building phase.

What is the market for the organic vegetables? The market is crucial when considering conversion and market research is essential. It is important to have identified potential markets prior to committing land to conversion. The mid-nineties saw many large arable and vegetable producers convert when the market appeared wide open and government grants for organic conversion first appeared. Subsequently the market has slowed and shows signs of becoming saturated, as produce from these conversions has reached the organic market and prices have dropped for many crops, such as potatoes and cabbage. However, there are still opportunities for certain crops, early and late in the season, and in the wholesale and direct-marketing sector. For new growers entering the market, it can be a lot of work establishing new markets. However, established growers have often been able to use their existing outlets for their produce and these have sometimes been the driver for conversion. In certain areas, producer co-ops exist, such as EOSTRE (based in East Anglia), and they can play a crucial role in developing the organic market, in areas such as public procurement of organic food, and for gaining access to larger markets that might otherwise not be available. Establishing direct-marketing schemes, such as box or community-supported agriculture schemes, is also very time consuming and there is now also a considerable amount of competition in some parts of the country. It is wise to start on a small scale and build up as skills are developed. It is generally true that a diversity of marketing options is beneficial to an organic business and will enable a larger proportion of the crops grown to be marketed.

How intensive is the farming system and what changes need to be made in order to be organic? Mixed farms have traditionally been the easiest and most readily converted, with the livestock and grassland providing the fertility for the vegetables. More recently there have been more stockless systems converting, relying primarily on grass/clover for fertility. The more widespread availability of green-waste compost should aid the sustainability and uptake of these systems.

What are the landscape characteristics of the farm? The shape and size of fields and their distribution is an important consideration. A ring-fenced farm is easier to convert than a farm that is fragmented. This is due to the issue of boundaries and potential risk of spray-drift. Effective windbreaks are essential for preventing spray-drift; where they do not exist or are deemed to be ineffective, then a 10 m buffer zone must be operated. This can have a big impact on the usable area of a field if surrounded by conventional arable land.

What is the condition of the land? High populations of perennial weeds such as creeping thistle (*Cirsium arvense*), couch grass (*Elytrigia repens*) and docks (*Rumex* spp.) can pose considerable challenges for organic conversion. A clean start is preferable and it is, of course, possible to use conventional methods of weed control, prior to conversion. Fallowing techniques, as outlined in Chapter 7, can be effective, particularly in dry conditions, but reduce opportunities for fertility building or cropping. High pressure from annual weeds, particularly on black fen soils or resulting from previous cropping or set-aside can also be challenging.

Are the soils suitable for vegetable production? While clay soils can be very fertile they do need careful management in terms of appropriate timings of cultivations (when not too wet). They are also slow to warm up in the spring and, therefore, not ideal for early production. In contrast, sandy soils, though easily cultivated and quick to warm up, can be very hungry for nutrients and water. The soil type can be important for crop quality, in terms of skin finish for root crops, when trying to meet supermarket specifications. The physical condition of the soil is another factor, as very poorly structured or compacted soil will take longer to recover under organic management and the effects are more visible and less easily remedied. Intensive prior use of the land for vegetable production can be a drawback if a legacy of soil-borne diseases or pest infestations has been left. White rot (*Sclerotium cepivorum*) in onions and club root (*Plasmodiophora brassicae*) of brassicas can remain a problem in the soil for twenty years or more.

Is the farm infrastructure suitable for an organic enterprise? The farm infrastructure is also important. The availability of irrigation is crucial for establishment of vegetable crops in most areas of the country, especially on lighter land. A good selection of farm buildings is useful, for packing and storage of vegetables, and if conventional production is to be maintained there needs to be clearly defined areas for each. There may need to be areas for manure storage or handling. If livestock are to be added to the system, adequate housing and secure fencing will be required.

Advice

When pondering the decision as to whether a farm system is suitable for conversion it is vitally important to get advice (*see also* Chapter 13). Whilst the details may vary from country to country, in the UK the first step is to contact the government-funded Organic Conversion Information Service (OCIS), which provides a telephone helpline service. Farmers and growers in any area of England can arrange for a free half-day visit and report by an adviser experienced in organic production and marketing. This can be followed by a full-day visit and expanded report by the advisor. The aim of the service is to provide impartial advice relevant to the business and the likely implications of converting to organic production. In England, the Organic Advisory Service (OAS), based at Elm Farm Research Centre (EFRC), provides this service, as does the Organic Centre Wales (OCW) in Wales. In Scotland and Northern Ireland the service is provided as a telephone helpline only (*see* Useful Addresses).

Careful study of the organic standards is essential and some certifying bodies have electronic versions that can be downloaded from their websites. Alternatively a charge may be made for a hard copy, which is deductible if you register for conversion.

Planning for Conversion

Conversion can be as much about learning and a change in attitude ('converting the space between the ears') as about physical changes to the land and farming system. Detailed financial budgeting and technical planning are, however, required, as changes to husbandry techniques, investment in equipment and infrastructure may be needed.

Gathering Information

Research and information gathering are an important part of planning for the conversion process (*see also* Chapter 13). Farmers in HDRA's Conversion to Organic Field Vegetable Production project spent between 420–520 hours gathering information during the conversion period. Visits to established organic farms and attendance at organic seminars and workshops is vital, to gain understanding of working organic systems. The SA, EFRC, HDRA, OCW and others all run full programmes of farm walks and events. In addition to the free advice from OCIS (*see* above), it can pay to take further advice, which is available from a number of advisory organizations, including Abacus, ADAS and EFRC. Conversion plans, feasibility studies and telephone support can be arranged. With the necessity for rotation, and especially if local markets are going to be supplied, it is usual that new crops will be grown or enterprises will be established. This in itself will be a learning curve and training may be needed.

Conversion Plan

Application needs to be made to a certifying body in order to begin the certification process. Certifying bodies usually require a conversion plan to be supplied with the application and this should normally cover the period of one complete rotation. This must include programmes for soil management, rotations, plant nutrition and pest, disease and weed control. Where applicable, programmes for grazing, grassland management, feeding, animal welfare and health, and environmental conservation are also needed. The purpose is to demonstrate that the applicant has a full understanding of the organic standards and farming system.

It may be necessary to decide whether to put all or part of a farm into conversion. In the UK, organic and conventional land are allowed on the same farm, as long as the area converted can be regarded as a sustainable unit in its own right, is clearly defined, and there are no opportunities for fraudulent switching of conventional and organic produce. In contrast, in parts of Europe it is necessary to make a commitment to convert the whole farm. It may be simpler, for record-keeping purposes, to create a separate holding for the organic enterprise. It should be borne in mind that it is prohibited to grow the same variety of conventional and organic crops in parallel.

One common mistake has been to put the least productive part of a farm into conversion, perhaps because it is not performing well in conventional production. Organic production, not surprisingly, does best on land that is in good heart, is easily workable and has few problems with weeds or pests and diseases: it follows that converting an unsuitable area can create problems from the outset.

It is common for arable and larger intensive vegetable farms to have staged conversions, whereas mixed farms with livestock often convert the whole farm at once. This is because livestock can be converted simultaneously, whereby the livestock (breeding animals only), pasturage and any land used for animal feed are converted at the same time. If the conversion is to be phased in stages, then it has to be decided which field, or fields, need to be entered first. The date of attaining organic status has to be considered here, which will be twenty-four months from registration and the last use of any prohibited materials. If possible this should be in early spring or before, because a late start in the first organic year, for example in June, can put pressure on the farming system, as many crops will need to be planted at the same time and this can have knock-on effects, with peaks of weed pressure and demands on labour and machinery.

The Conversion Period

Conversion Strategies

During the conversion period the land has to be managed according to the organic standards and documentation and record keeping must reflect this. Different strategies can be adopted during this time, which will partly depend on the market for in-conversion produce and partly on the type of farming system that is being converted. After at least twelve months have elapsed from the start of conversion, crops harvested can be sold as 'in-conversion'. However, there isn't really a market for in-conversion vegetables as such, although direct marketing through farmers' markets and box schemes, where the seller can communicate directly with the consumers, may be a better outlet. For mixed arable conversions there may be the option of growing in-conversion cereals, but the state of the market should be checked out first with Organic Grain Link or the Organic Arable Marketing Group.

Fertility Building

Once the plunge has been taken and the land has been registered with a certifying body it is important not just to sit back and wait for the two years to pass. The conversion period is a good opportunity to begin to build soil fertility and encourage the build-up of biological life in the soil, especially where the land has been under continuous arable cropping. Fertility-building crops should be considered as important crops in their own right and need to be nurtured in the same way as any other, needing good conditions for establishment, irrigation during dry weather and proper management. Although the use of cut and mulch grass/clovers or vetches is common (*see* Chapter 4) in organic systems, they can be a new technique for many arable farmers. Manures and composts can also help and new systems for dealing with them might need to be implemented. Regular mowing of grass leys will be needed, cutting when no higher than an average knee-height down to ankle length. The idea is to avoid too much bulky material being cut, which can form a thatch preventing re-growth of the ley. Mowing will select against weeds and prevent seeding of the grass, clover and weeds.

Farm Enterprises and Cropping

It is likely that changes will be required to the farming system, but the extent of those will depend on how specialist, intensive or mixed the conventional farming system is. Usually the system will become more diverse and complex and, typically, management time will increase.

For specialist vegetable farms converting to organic, the principal change to the farming system will be fertility management (*see* above). Fertility building will be needed and the growing of grass and clover or other legumes may well be unfamiliar to the conventional grower. If the farm has already been supplying local or direct markets, such as farmers' markets or farm shops, then it will already be growing a variety of different crops and may need fewer changes. The more specialist the production, the more changes will be needed.

Large arable/vegetable farms had been slower to convert than other farm types until the boom of the organic market and the introduction of conversion grants in the late-nineties. A HDRA study has observed that conversion has been technically possible on this farm type, but generally has required large investments in machinery, especially for weed control to reduce hand labour. The market outlets are mostly to the packers and supermarkets and the scale of production means that risks are high. The same study showed that arable farms converting to include vegetables tended to grow fewer cereals, grow more fertility-building grass leys and grow more vegetables after conversion. These large-scale arable conversions used a staged conversion, typically over five or six years. Fewer break crops, such as sugar beet and oil-seed rape, were grown, as there is no organic market, at present, for some of these crops that are grown conventionally. A wider variety of vegetables are usually grown, some of which the farm may be unfamiliar with. Livestock often fit well into such an organic system, but the economics and current market situation should be examined closely if livestock are to be re-introduced, especially if skills and infrastructure are lacking on the farm.

The mixed-farming system with livestock, fodder and arable crops will generally involve less radical changes during conversion. The importance of clover in leys may need to be rediscovered but there is more likely to be an existing rotation. Manure-handling systems might have to be changed, but farmyard manure, together with grass/clover leys can provide the basis of fertility. Vegetables can rotate around the farm, with grassland and other crops, and a long rotation can minimize any problems from pests or diseases and weeds. Traditionally, many more farms of this type have converted to organic, particularly in the West of England and Wales. The soils of these farms may be heavier and less than ideal for vegetable production. Many traditional grassland farms have soils that are low in phosphorus and potassium but are able to support good crops. If the farms have not grown vegetables before, then investment in machinery, irrigation and packing facilities may be needed. Conversion of these farms tends to be shorter, as they often undertake the simultaneous conversion of livestock. This whole farm conversion can lead to a drop in income during the conversion period, though livestock will still be bringing in some income.

Chapter 3

Crops, Seed and Varieties

A wide range of vegetable crops can potentially be grown in organic vegetable-production systems. Within each crop type, a diverse and often bewildering array of varieties is usually available. Choice of which variety to use and the variety that will perform well in any given situation will depend on many factors. Some of these are under the control of the farmer or grower and others are not. Factors that can be controlled to some extent are those like planting time, soil condition or marketing outlets, whilst those that cannot include factors such as availability of organic seed, soil type, rainfall and market quality requirements.

Obviously, crop and variety choice will ultimately be a compromise between conflicting needs within the farm system and some of these trade-offs are briefly discussed in this chapter as an overview of the main vegetable crops and varieties – a more detailed discussion of the crop types is left to Chapter 12. Choice of cultivar can also be helped by variety-trialling information and this is also discussed. The use of organic seed and organic plant-raising media and techniques is a necessity for production under organic standards and the major issues are briefly touched upon.

VARIETY TRAITS FOR ORGANIC PRODUCTION

Organic farmers and growers largely depend on varieties supplied by conventional plant-breeders and the conventional (non-organic) seed industry. Whilst many of these varieties are excellent, it should be borne in mind that the majority have been bred to perform well in farming systems where artificial fertilizers and pesticides are routinely used. This is obviously not typical of organic farming systems and although varieties often perform well in both conventional and organic systems this is not always the case. In general, organic farmers should look for 'resilient' varieties that are able to provide a stable yield over a range of conditions. Growing varieties that yield well only under more exacting and protected conditions obviously implies a greater risk to yield and performance when conditions are adverse, such as under high disease pressure or low soil fertility. Although it is difficult to generalize across crop types, especially with crop types as diverse as vegetables, it is possible to sketch the general traits that can be desirable in organic vegetable varieties. Any specific traits of importance are mentioned in the discussion on specific crop types.

Generally, organic varieties should be adapted to organic soil-management regimes (*see* Chapters 4 and 7). This implies a lower level of inputs, potentially spread over a longer time period, which, in turn, means varieties with better and more well-developed root systems and roots capable of forming interactions with soil micro-organisms. Plants should be able to tolerate at least some periods of nutrient stress. It can be difficult to pick out varieties with these 'hidden' characteristics but, at the least, varieties should be selected under organic growing conditions, in the appropriate place in the rotation, to evaluate this characteristic. Some crop types and/or varieties will also be better adapted to production in certain soil types.

Weeds are an important constraint in many vegetable-production systems (*see* Chapter 5) and varieties that consistently suppress weeds are generally desirable. Organic varieties should, therefore, show rapid early growth and the ability to cover the soil and shade it to outcompete weeds for light at as early a stage as possible in the crop cycle. Varieties are well known to differ in both architecture and competitive ability against weeds and those that outcompete weeds for light, nutrients and water are preferred. Crop varieties, often those with erect foliage, that can tolerate some degree of mechanical or hand weeding are also to be preferred. Some crop varieties may also produce inhibitory substances or allelochemicals that prevent weeds from germinating or developing, although information about this aspect is limited.

Pest and disease resistance is also an important consideration in choosing organic vegetable varieties (*see* Chapter 6). In general, more resistant varieties are to be preferred, with resistance to air-borne diseases being uppermost rather than soil-borne diseases, which can be managed by rotation. Plant architecture can also be important in reducing susceptibility to disease (such as a more open growth habit that avoids humidity building up in the canopy) as can other traits, such as the ability to grow as intercrops or in mixtures. Quick establishment and plant development followed by early maturity will often help to avoid disease pressure, as exposure to pathogens is limited.

Organic varieties should also attain the product quality necessary for marketing needs (*see* Chapter 10). Traits that are important to organic consumers vary from the stringent demands of packers and supermarkets

Evaluating lettuce varieties for organic production

to the more tolerant attitude of box-scheme customers. Traits such as size, shape and taste will be important to varying degrees and it will be necessary to compromise between characteristics in order to meet the demands of both the production system and the market. However, and on the whole, consumers do expect organic produce to taste better and this should be an important consideration.

VEGETABLE CROPS AND VARIETIES

Apart from the general requirements for organic variety characteristics, each crop will have specific needs depending on the crop type. Considerations will include regionality, seasonality and specialist or niche crops. The main vegetable types are shown in Table 3.1 and the most important plant families and crop types are briefly discussed below. Many of the themes are further developed in Chapter 12, where the management methods for the various crop types are discussed in more detail.

Alliums (Amaryllidaceae)

Alliums are normally grown for their bulbs or stems. They are not generally regarded as nutrient-demanding vegetables and can generally be grown later in the rotation, after more hungry crops. In a rotation, alliums should not be grown more than one year in four and, where soil-borne diseases are a problem, rotations should be extended beyond this.

Table 3.1 List of major vegetable crop types grown in organic systems (by type and botanical family)

Vegetable type	Family	Crop
Leafy Salad	Cruciferae (Brassicaceae)	Chinese Cabbage
		Mustards
		Rocket
		Mizuna
		Pak Choi
	Chenopodiaceae	Leaf Beet
		Spinach
		Swiss Chard
	Compositae (Asteraceae)	Chicory
		Endive
		Lettuce (Batavian, Butterhead, Cos, Little Gem, Iceberg, Looseleaf)
Leaf/Flower	Chenopodiaceae	Chard
		Leaf Beet
		Spinach
	Cruciferae (Brassicaceae)	Cauliflower
		Brussels Sprouts
		Cabbage (Red, White, Savoy)
		Calabrese/Broccoli
		Kale
		Purple-sprouting Broccoli
Fruit	Cucurbitaceae	Cucumbers
		Melons
		Pumpkins
		Summer Squashes (Butternut, Courgettes, Marrows)
		Winter Squashes
	Solanaceae	Aubergine
		Peppers
		Tomatoes
Pod or Grain-Bearing	Gramineae	Sweetcorn
	Leguminosae (Fabaceae)	Beans (French Dwarf or Climbing)
		Beans (Runner)
		Beans (Field or Broad)

Table 3.1 (continued)

Vegetable type	Family	Crop
		Peas (Vining, Podding)
		Mangetout
Bulb/Stem	Alliums (Amaryllidaceae)	Onions (Red, White, Spring)
		Shallots
		Leeks
		Garlic
	Cruciferae (Brassicaceae)	Kohlrabi
		Radish
	Umbelliferae (Apiaceae)	Celery
		Celeriac
Root	Chenopodiaceae	Beetroot
	Cruciferae (Brassicaceae)	Swede
		Turnip
	Solanaceae	Potato
	Umbelliferae (Apiaceae)	Carrot
		Parsnip
Perennial	Compositae (Asteraceae)	Jerusalem Artichoke
		Globe Artichoke
	Liliaceae	Asparagus

Types and Varieties

A range of allium crop types are routinely grown in temperate vegetable productions systems. The most common are mentioned below.

Onions: two main types are usually recognized, red and white bulb-onions, although other types such as spring onions and shallots can also be grown. Onions are often grown from sets or module transplants but, with the use of sophisticated weeding equipment, direct drilling is becoming an increasingly popular choice. The major issues to consider when choosing varieties are resistance to downy mildew (*Peronospora destructor*) and the ability to withstand the weeding regime (as onion crops are likely to offer little competition to weeds). Weeding regimes can include flaming post-emergence, but they are more likely to be largely cultural and mechanical. Depending on market, storage may be important after harvest and, where it is, varieties with good storage ability should be chosen.

Leeks: in general, the requirements of organic leek growers match those of conventional growers. The main problems encountered in trialling have been weed control and rust infection. A range of varieties is available to suit different markets, so these should be identified before selecting the variety. Optimum production periods may also vary between varieties, depending on the time of harvesting (autumn or spring periods), and this should also be taken into account. Open-pollinated (OP) varieties have been traditionally used by organic growers, but current work is comparing hybrid to OP varieties in organic systems. No varieties immune to rust have been identified, but some appear less susceptible than others. Although varieties with spreading foliage might offer better weed smother, organic growers might prefer more erect varieties that allow more frequent mechanical weeding without damaging the plants.

Garlic: is a high-value crop widely grown. For direct-marketing schemes it can be stored and sold over an extended period of time.

Soil Fertility and Rotation

Good rotation is essential for minimizing build up of pests and soil-borne disease problems. Alliums must not, under organic standards, return to the same land before a period of three seasons has elapsed. If it is possible to extend this further then that could be beneficial. The standards allow successional cropping of the same family in the same year (for example, spring onions), but if it can be avoided, for the reasons stated above, then do so. Onions and leeks are not too nutritionally demanding and can come later in the rotation, using residual fertility from previous crops. In any case, it is best that they don't follow a grass/clover ley, due to risk of wireworm. Where cover crops are used, in particular before leeks, it is best to avoid wheat and rye, which provide excellent over-wintering sites for thrips. Onions may follow cereals, green manures or certain vegetable crops in the rotation. They should not follow crops attacked by the same strain of stem and bulb eelworm (namely, oats, clover, lucerne, sugar beet, leek, carrot, parsnip, pea, broad bean, dwarf bean, rhubarb, strawberry, daffodil or tulip). Potatoes can precede onions, but volunteer potatoes can become a serious weed problem.

Weeds

None of the alliums are particularly competitive and controlling weeds is likely to be the biggest challenge. They germinate and grow slowly; they generally have few leaves, an upright habit and do not cover the ground well. Weeds can, therefore, germinate over a long period. Weeds can be more of a problem in crops grown from seed than in transplanted or vegetatively propagated crops. At crop harvest, weeds foul undercutting and lifting machinery and prevent onion bulbs from drying in the windrow. The main methods of weed management are cultural, mechanical and thermal. Cultural controls include choice of growing system (for example, sets or modules), time of planting and spacing. Mechanical control includes harrowing and hoeing, while thermal control involves flame weeding to control small seedling weeds. The success of these methods depends on timing, on weather and soil conditions and on the composition and density of the weed population. Alliums such as bulb onions can be grown through sheeted mulches, such as plastic or starch polymers, which should prevent weed emergence across the entire cropping bed.

Pests and Diseases

Allium crops naturally tend to share a range of pest problems. Rabbits can be a surprisingly devastating pest of allium crops, as they will graze happily on the young shoots of seedlings and transplants. Rabbit fencing or electric netting around the plot is necessary until the crop is well established. Of the insect pests, onion fly (*Delia antiqua*) maggots feed on the bulbs or leaves of onions, whilst thrips (*Thrips tabaci*) are a sap-feeding and troublesome insect pest on leeks. Leek moths (*Acrolepiopsis assectella*) occasionally cause serious problems on leeks, depending on the season. Good cultural management and general sanitation should help to reduce damage due to these pests. Cutworms, the caterpillar stages of several moth species, feed on young allium stems just above ground level and can burrow into bulbs of older plants. They are most active in mid-June and can be more serious on weedy land. Warning services forecasting cutworm activity are available and irrigation usually prevents serious damage, as the moths and young larvae die in wet soil. Similarly, wireworms (*Agriotes* spp.) are the larvae of the click beetle and can be a problem after ploughing long-term grass leys, burrowing into the roots and stems. Consolidation of the seedbed can reduce their activity.

Stem nematode (*Ditylenchus dipsaci*) has a wide host-range and affects alliums, where it causes twisting, distortion and discoloration of stems and foliage, and distorted and cracked bulbs. Infection can be introduced by contaminated seed or planting material, so it is vital to obtain clean seed. Infected onion sets tend to be soft, shrunken and discoloured (dark brown) near the neck of the plant and are lighter in weight. Severe infestations tend to be worst in wet-weather periods. Rotation is important, as nematodes can survive in moist soil for about a year. The use of compost has been shown to have a suppressing effect on nematodes in some situations.

White rot (*Sclerotium cepivorum*) affects all alliums and is the most damaging disease. The first signs are leaves yellowing and wilting. When the plants are pulled up they will have a fluffy, white rot starting from the base. Onions are usually affected more seriously than leeks or garlic. It is a very persistent soil-borne fungus and can remain viable in the soil for up to twenty years. It can be spread by cultivations and any infected plants should be removed and burnt if practicable. A long rotation should be practised.

Downy mildew (*Peronospora destructor*) is principally a disease of onions, manifesting itself as purplish-grey spores on large leaf blotches, which may be at the leaf tip. Given the right conditions (warm and humid), it can spread very rapidly through the crop, causing collapse of foliage and secondary sooty moulds. If infection occurs early, which is more likely after a mild winter, it can seriously reduce yields and result in soft, immature bulbs. Sheltered sites should be avoided and successive plantings should be isolated from each other.

Rust (*Puccinia allii*) is primarily a problem with leeks and garlic. Small, bright orange pustules appear from July onwards in warm and wet conditions, reducing marketable quality and causing death of foliage when severe. Degrees of varietal resistance exist and careful choice is needed. Isolate successive plantings if possible, to reduce risk of spread.

Damage from white tip (*Phtophthora porri*) consists of water-soaked white lesions at the tips of leaves and dieback in late summer or autumn, particularly of leeks. It is a soil-borne disease that can survive for several years. Rotations should be extended if it is a problem.

Neck rot (*Botrytis allii*) is a disease that occurs in the field and in storage. It is a seed-borne fungus, but can also be carried in sets of onions and shallots. Although it can be present at harvest, it is more often observed in storage, when a soft brown rot develops from the neck. Topping of the foliage prior to harvest can allow an entry point for the disease. The main control method is to prevent the fungus growing down the neck of the bulb after topping. Topping to leave a neck of about 80mm and rapid curing is advised.

Botrytis leaf spot and leaf rot (*Botrytis squamosa* and *Botrytis cinerea*) cause small, white spots on the leaves, with a water-soaked margin or halo, during wet or humid periods. It is particularly a problem with salad onions, where the blemishes can make them unmarketable. General crop-sanitation measures together with rotation should help to reduce incidence of this disease.

Chenopodiaceae (Beets)

The beets, chard and spinach are salad leaf crops from the Chenopodiaceae family. They should be placed together with beetroot, also in this family, in a rotation, although organic standards do not restrict the time in which this family of crops can be returned to the same location.

Types and Varieties

Beetroot: although a biennial, beets are generally grown as annuals and are likely to seed if left to grow. Beets are mainly grown for their edible roots, which come in a wide range of shapes and colours. Red, round types are generally preferred and grown in the UK but other shapes and colours can include long, oblong and cylindrical and they can vary from dark red, dark purple-red to yellow or white.

Leaf Beet, Chard, Swiss Chard, Spinach Beet: are all variants of common beet (*Beta vulgaris*), which includes beetroot (*see* above). Once again, they are biennials that are grown as annuals and readily go to seed if left. A range of varieties is available, running from green-leaved ones through to varieties with ornamental foliage (greenish-white or red foliage). The stems are also edible and can be quite broad in some varieties. They will generally come back when cut, but will tend not to thrive in hot weather. The stems and leaves of chard are usually cooked before being eaten.

Spinach: there are various types of spinach, ranging from baby leaf and flat types to crinkled types. Spinach can generally be grown as spring, summer and winter varieties and is a handy green for including in salad packs and for direct sales.

Soil Fertility and Rotation

Beets prefer a cool, moist soil and might benefit from mulching or irrigation to maintain these conditions. If scab is a problem, as long an interval as possible should be left between crops in the rotation.

Weeds

Beets are intolerant of, but not very competitive with weeds, especially if the leaves are continually cut back. Management requires adequate field preparation and stale seedbeds to reduce the weed population from the outset. Hand hoeing may be necessary to avoid damaging the root crop.

Pests and Diseases

These plants generally suffer from few insect pests. In contrast, a number of foliage diseases are commonly observed, though rarely serious. Ramularia leaf spot (*Ramularia beticola* or *Cercospora beticola*) is widespread and can generally be managed by removing affected leaves. Some powdery mildew (*Erisyphe betae*) causes a blemish to foliage, as can rust (*Uromyces betae*). Normal crop husbandry and hygiene measures should be sufficient to manage these diseases. Scab (*Streptomyces* spp.) can be a problem on the roots of beetroot, causing unsightly swollen outgrowths. Rotation and use of green manures may help to alleviate this problem. Viruses, such as cucumber mosaic virus (CMV) and beet yellows, can affect beets and spinach. The diseases should be managed by eliminating dumps and volunteer plants. Beets are known for being sensitive to boron deficiency, which can cause blackspot, sickly growth and poor taste, depending on the circumstances.

Compositae (Asteraceae)

The Compositae is another large plant family, of which there are many native examples in the UK. This family group is mainly noted for leafy salad crops, like lettuce, endive and chicory, but crops such as globe artichokes also belong to this group. They share some similarities in cultivation techniques, as well as pests and diseases, but, in practice, represent a diverse group.

Types and Varieties

Compositae are mainly grown for their leaves, which are used in the increasingly popular salad packs or as stand-alone heads. The artichokes, which are not very closely related to the leaf types or to each other, are perennial vegetables that are often a useful addition to a range in the direct-sales market.

Chicory: will grow under a wide range of conditions and can be a useful addition to the range of leafy vegetables grown on a holding. It can be cut as whole heads or as cut-and-come-again leaves. A moderate number of varieties of different types are available for organic growers.

Lettuce: a large number of lettuce varieties are available as organic seed and, although organic lettuce trialling has been carried out since 1991, only a limited number of the total available has been tested. All types of lettuce have been trialled, including crisphead, butterhead, cos and leaf types. Important varietal characteristics for successful organic lettuce are vigour and disease resistance, mainly to downy mildew. Fortunately, there is a wide range of genetic resistance available in lettuce varieties, including downy mildew resistance, aphid resistance (both root and leaf) and LMV (virus) resistance. Many of the varieties available are continental types, such as butterhead and Batavian lettuce. Little gem (cos) types are often popular with organic growers. In trials, crisphead varieties have often had difficulty achieving head weights under organic conditions. Due to the wide range of varieties available as organic seed, it is likely that growers will have to experiment with a range of varieties and types to find those best suited to their own production systems and markets, although the information available from variety-testing programmes will certainly help make the initial selections.

Endive: is an easily grown crop, which is mainly grown for leaves and as a cut-and-come-again crop. Varieties are available for summer growing and over-wintering, including some frost-tolerant varieties.

Jerusalem Artichokes: a perennial vegetable closely related to the sunflower, with which it shares a similar growth habit. They are easy to grow and, once planted, come back each year. They may have a limited appeal to all but dedicated box fans.

Globe Artichokes: also a perennial and specialist crop, normally grown for its flower heads and can remain productive for about three years.

Soil Fertility and Rotation

There are no limits to the length of time between crops of this family but, where soil-borne diseases are troublesome (for example lettuce), two or more years should be left between crops. More details on the individual crop types are given in Chapter 12.

Weeds

Weeds are best managed with cultural measures from the outset, for example by using stale seedbeds. Steerage hoes can be used when the crops are still small. Hand weeding may cause damage to the more delicate crops, especially lettuce, and should be avoided if possible once the heads are developing. Sheeted mulches can be used in these crops.

Pests and Diseases

A range of pest and diseases are likely to be encountered on these crops – pests and diseases are discussed in more detail under the specific crops in Chapter 12. Most of the crops, apart from lettuce, suffer little from pest attack.

Aphids (various species) are the main insect pests of lettuce, with the other types suffering little pest damage in general. However, consumer tolerance for even one insect in a head is very low, despite the fact that they do no damage at this level. Aphids can also transmit viruses between plants and between weeds and plants. Pests such as rabbits and slugs can also be a problem in lettuce in some circumstances.

Foliar diseases can be serious, as they affect the part of the plant that is sold. Downy mildew (*Bremia lactucae*) is commonly regarded as the most problematic disease in lettuce at the seedling/transplant stage and also on the crop towards the end of the season. Resistant varieties are available and general crop hygiene to destroy debris is necessary, especially if successional crops are being grown. Lettuce ringspot (*Macrodochium panattonianum*), a fungus, can attack outdoor crops of chicory, endive and lettuce in cold, wet weather. Sclerotinina disease (*Sclerotinia sclerotium* or *S. minor*) can also attack the base of the plants and is most often observed under cool, damp conditions. This disease is best managed by crop rotation, leaving at least three years between susceptible crops and taking care to control weeds from the same family that can also succumb to the disease.

Cruciferae (Brassicaceae)

Brassicas are represented by a range of vegetable types. These include freshly eaten leafy salad vegetables as well as cooking vegetables. All parts of the plant have been selected for development at one time or another, including the roots (for example swede), stem (for example kholrabi), leaves (for example kale), axillary buds (for example Brussels sprouts) and flowers (for example broccoli). Seasonality is a major factor in growing brassicas. Brassicas are nutrient-demanding and should be placed accordingly in a rotation. Due to pressure of pests and diseases they should not be grown in the same place for more than one year in four and typically benefit from longer rotations than this. Traditionally they have been grown on alkaline soils – often because of clubroot (*see* below).

Types and Varieties

A wide range of different brassica types exists and one type or another can potentially be grown throughout the year. Some types are the traditional 'greens', whilst others are being increasingly grown as leafy salad vegetables. Some of the more commonly grown types are listed below (more information is also given in Chapter 12). Within each type a more or less wide variety choice exists. Seasonality, colour and shape can be critical for marketing purposes and varieties should be chosen with this in mind. Resistance to pests and diseases can be a benefit where it exists, as it can help to reduce the use of fleeces or other labour-intensive methods.

Brussels Sprouts: generally thought of as a difficult crop to grow organically, partly as a consequence of the length of time the crop remains in the field and the peak Christmas demand. Resistance to aphids is desirable, especially on mid-season varieties, as well as good height and vigour. In trials, differences in susceptibility to diseases have been observed and less-susceptible varieties should generally be chosen if possible.

Cabbage: a wide range of cabbage types exists and choice of variety will depend largely on the market, especially as regards appearance of cabbage (red, white, green, savoy and so on) and timing (spring, summer, autumn or winter). In trials, aphid infestation has proved important in choosing between varieties and, in general, smooth varieties are less affected than blistered, red less than green or white types and early maturing varieties less than later maturing ones.

Calabrese/Broccoli: a range of summer, autumn and winter varieties is available. Many of the most popular varieties are hybrids, which often do not achieve high crown weights under organic conditions. This crop is generally competitive with weeds and relatively trouble free in this respect, but can suffer root-fly damage when the plants are small.

Cauliflower: a range of cauliflower varieties that mature in summer, autumn and winter is available. Many of the varieties seem to mature at a slower rate under organic conditions. Aphid and caterpillar damage have been major problems and cabbage root fly can be a problem in young plants. Diseases such as ringspot can also be a problem if the crop stands over winter, as can diseases like *Alternaria* that affect the curd.

Chinese Cabbage: a leafy green vegetable becoming increasingly popular as a substitute for crops like lettuce, especially over the winter months, with many different types now becoming available.

Kale: popular as a winter greens crop, albeit generally cooked. Many varieties are currently available.

Oriental Vegetables: becoming increasingly popular as winter greens and for salad and stir-fry packs. They are also used in 'baby leaf' production. The list of types and varieties is continually growing and includes crops like Mibuna, Mizuna, Oriental Mustards and Pak Choi. Although they are brassicas, they tend to suffer less from the typical brassica diseases, possibly because they tend to be grown out of season or under cover, but flea beetles can be a problem.

Turnips and Swedes: are featured in many organic rotations. Cabbage root fly, slugs and powdery mildew have all been noted as problems at one time or another.

Raising and Growing Brassicas

With the exception of oriental greens, turnips and swedes, most brassicas are sown into modules or blocks and then transplanted. They are often bought into a holding as transplants. Generally they are sown one seed per module or block and covered until germination. Most brassicas germinate quickly – within two days – and it is important that the cover is removed as soon as the first signs of germination appear, otherwise 'leggy' transplants will be produced. Brassicas will be ready to transplant when they have attained 4–5 true leaves, which will take between 4–8 weeks depending on the time of year. Shortly before transplanting, seedlings can be moved to an outside standing area to 'harden off'.

Seedlings are most commonly transplanted into beds. Spacing depends on the type of brassica (*see* Chapter 12). Brassica seedlings are generally hardy and will withstand some degree of transplant shock.

Soil Fertility and Rotation

It is not advisable to grow brassicas for more than one year in five in a rotation, to prevent the build up of diseases, particularly clubroot. Brassicas are generally a nutrient-hungry crop, so are most commonly placed directly after the fertility-building phase in a rotation or at least in the first half of the rotation. Crops with inadequate nitrogen will produce stunted plants with pale leaves. This will result in small frame size to support growth of the head, so yield and quality are reduced. Reasonably high levels of organic matter are also required. On soils with very high levels of organic matter, such as peat fenland, manganese deficiency may be observed as a bluish tinge. Applying manganese supplements under an organic derogation may rectify this condition. They do better in a neutral-to-alkaline soil and often this is a good place to add lime to a rotation, should soil tests indicate that it is necessary or if the farm has a history of clubroot disease (*see* below).

It is essential that crops are supplied with adequate moisture, especially during critical stages such as head formation in cauliflower or calabrese. Drip irrigation will reduce the chances of fungal infection by applying water directly to the roots, but is far more expensive to set up than other forms of irrigation.

Weeds

Many of the brassica types are raised and transplanted as modules and this can help reduce weed problems if the ground is adequately prepared before planting. Many of the brassicas are leafy and are able to outcompete weeds, given a good start. They can often be grown without the need for large amounts of hand weeding. Once the transplant has established in the soil, harrows can be used quite aggressively and repeatedly to remove weeds across the entire bed. Brush weeders can be used on lighter soils to remove or cover small weed seedlings close to the transplants when plants are still at a small stage. Once plants are more robust, finger weeders will achieve good intra-row weed control, on lighter soils. On heavier soils, steerage hoes can be used to cut through and bury weeds within the row and cover those between the rows. Later in the season, many brassicas achieve adequate canopy cover to suppress weeds.

Pests and Diseases

The various types of brassicas tend to share a common set of pests and diseases and the importance of the pest will depend on which part of the plant is attacked in relation to the harvested or marketed part. There are many pests that attack brassica crops.

Flea beetles (*Phyllotreta* spp.) are small, shiny beetles up to 3mm long that are easily identified, as they jump like fleas when disturbed. Early in the season, flea beetles can cause widespread damage by making holes in the leaves and can even kill off small plants. It may be advisable to cover early crops with fleece during April and May, when the beetles are most active. If covering the crop, it is important to do this immediately after the crop is transplanted, as covering it once the flea beetles are observed is too late. Later in the season, crops will grow through a small amount of damage.

Of the cabbage white butterfly (*Pieris* spp.) family, the small cabbage white (*Pieris rape*) is the most damaging species. It is present in two to three generations per year, from June to late autumn, and forms holes in the leaves. The centre of plants may be attacked and fouled with excrement. In extreme cases crops may need to be sprayed with the agent *Bacillus thuringiensis* (Dipel). Crops destined for supermarkets may need more frequent doses, as tolerances for caterpillars and their droppings in the head are very low. The large white (*Pieris brassicae*) is also very common and can also do considerable damage, especially when large numbers of caterpillars are present. They usually skeletonize the outer leaves. Covering the crop with mesh is another option for protection.

Diamondback moth (*Plutella xylostella*) caterpillars are generally found feeding on the lower leaves and sometimes in the hearts of brassicas. They are capable of many generations in a year and can build up to quite high populations, although the initial population establishes from immigrants from the continent.

Several generations of cabbage root fly (*Delia radicum*) can take place from April–May onwards until August–September. The adults have a similar appearance to ordinary houseflies and lay eggs at the base of the plant. White maggot-like larvae hatch after about six days and can cause extensive damage to the plant by burrowing into the roots and the stem. The damage is most serious in young transplants, where the larvae can kill off plants. In older plants, damage to the root system can reduce yield, especially in dry conditions. The most common control for cabbage root fly is covering with fleece, which should be done immediately after transplanting. The best time for removal can be judged using a forecasting system, which predicts the peak times for egg laying. When deciding to use fleece, the considerable cost of material, and applying and removing it for weeding operations, must be balanced against the economic loss from the pest. If losses are only low, then the cost of using fleece will not be justified.

The two most common types of aphid found on brassicas are the peach potato aphid (*Myzus persicae*) and cabbage aphid (*Brevicoryne brassicae*). The cabbage aphid has the potential to do most harm, causing serious feeding damage, distorting and fouling of the leaves. It is easily recognizable by its

The large cabbage white caterpillar is often seen on brassicas during the summer

mealy, grey, waxy appearance and usually forms small colonies under the leaves. Peaks in population occur around September. The peach potato aphid is less likely to do direct damage to the crop. It can be pale- to dark-green and numbers usually reach peaks in mid- to late-July. The most favoured method of controlling aphids in organic systems is to build up populations of predators, such as hoverflies, ladybirds and lacewings, to control aphid numbers. Many of the adult forms of predators feed on pollen and nectar, so can be attracted into the crop by allowing wild flowers to grow around field margins or by planting beneficial strips of *Phacelia* or umbelliferous species. Covering the crop with mesh will exclude aphids, although this is an expensive measure both in terms of material and the extra labour involved in removing it for weeding operations. As a last resort, crops can be sprayed with a potassium soap to control aphids, although this is non-specific, so will also kill off predators.

Birds, such as pigeons (*Columba palumbus*) and crows (*Corvus* spp.), can cause a considerable amount of damage. Crows are keen on systematically pulling up large numbers of recently transplanted seedlings, which can drastically reduce plant population. Pigeons can also strip the leaves of young seedlings, killing the plants. Humming tape and bird scarers can be employed but, in extreme cases, it may be necessary to cover the crop with fleece for a few weeks, until it has had a chance to establish and can withstand a certain amount of bird damage.

Brassicas can succumb to a wide range of diseases but, in the majority of cases, especially in established crops, these remain at low levels, generally affecting older leaves, and do not cause economic damage to the crop.

Damping off is normally first noticed when seedlings collapse and die from the base of the stem. Damping off can cause large losses of seedlings at an early stage of the crop. It is most commonly caused by infection with the soil-borne fungi *Pythium ultimum* or *Rhizoctonia solani*, which can come from a range of sources, including the growing medium, contaminated water or dirty tools. It is favoured by damp soil conditions, so symptoms can be reduced by ensuring good drainage and ventilation.

Downy mildew (*Peronospora parasitica*) is frequently encountered at both the transplant-raising stage and in the field. It is most serious at the plant-raising stage, where it can kill large numbers of seedlings. It is observed as yellow patches on the leaves with white mould growing on the underside. It is especially troublesome if conditions are cool and moist. It is best avoided by ensuring good ventilation in the glasshouse and arranging the watering regime so that the foliage does not remain moist for long periods of time (namely, avoid overhead watering in the evening).

Black rot (*Xanthomonas campestris*) is a bacterial infection that appears as pale-yellow 'V'-shaped lesions at the leaf margins. It can kill off entire leaves on younger plants, but some infection can be tolerated on older plants. Outer leaves can be stripped off at harvest to remove infected leaves, although this may result in loss of yield if infection has progressed too far. Infection can be seed borne or through crop residues, so good sanitation, such as using clean seed and ploughing in crop remains, is essential.

Dark leaf-spot (*Alternaria brassicae, Alternaria brassicicola*) is a fungal disease causing small black spots that increase in size and coalesce to form larger target spots. It is seed borne but can also be carried by cruciferous weeds, such as charlock. At low levels it is of little consequence in crops where the outer leaves can be removed, but it may spoil the appearance and reduce the saleability of Brussels sprouts or the heads of cauliflowers. It is best avoided by using clean seed and ploughing in crop residues.

Ring spot (*Mycosphaerella brassicicola*) fungal disease causes large, grey lesions on the leaves and severe infections can defoliate plants. It is prevalent in wet conditions, especially in over-wintered crops, such as winter cauliflowers or Brussels sprouts. Again, good crop-sanitation measures, such as ploughing in crop residues and avoiding year-round production of brassicas, can reduce the chances of infection.

Clubroot (*Plasmodiophora brassicae*) is familiar to many growers from the characteristic gall-shaped roots resulting from this fungal disease. The damage it causes to the root system can result in plant death or serious reduction in yield. Once soil is infected, the disease can persist for many years, so it is prudent to take precautionary measures to avoid infection being introduced through infected plants, machinery or composts. The disease is inhibited by a pH of seven or above, so application of limestone to acidic soils before growing brassicas is advisable.

Cucurbitaceae (Squashes)

The Cucurbitaceae family is a diverse group of plants whose fruits can add a splash of colour or diversity to a vegetable-growing enterprise. Many of the domesticated plants derive from South America. They have become more popular again as tastes change and their culinary versatility is appreciated. They are mainly used for their fruits but, in some cases, the leaves and flowers can also be eaten.

Types and Varieties

A range of types exists and, within these types, many varieties are offered in seed catalogues. Further details are given in Chapter 12, but the main types are briefly described below.

Cucumbers: are an indoor or protected speciality crop in the UK.

Melons: are also often seen as a specialist crop, as they need high temperatures to develop, and are therefore best grown under protection.

Pumpkins: are usually aimed at the Halloween market but can also be stored for a long time, under the right conditions. A moderate range of varieties is offered to organic growers.

Summer Squashes: including courgettes and marrows, are all members of the cucurbit species (*Cucurbita pepo*). Summer squashes tend to be determinate types and they are used in the immature stage. A wide range of varieties is generally offered in seed catalogues, many of them organic (*see also* Chapter 12).

Winter Squashes: such as pumpkins, butternuts and gourds, are indeterminate in habit and will trail vigorously. Most winter squashes are either *Cucurbita moschata* or *maxima* species. Others are members of the *pepo* species (for example acorn squash and spaghetti squash). Once again, a wide range of varieties is generally available for organic growers.

Soil Fertility and Rotation

Cucurbits generally prefer well-drained and fertile soils. They are fairly nutrient-demanding (especially cucumbers) and should be placed accordingly in a rotation. Many will benefit from irrigation, especially when the fruit is forming.

Weeds

Outdoor squashes are generally planted into black, plastic mulch, which can help to advance the crop but also controls weeds. Hand hoeing might also be necessary in some crops, depending on circumstances. Indoor crops will need similar weeding care to all protected crops.

Pests and Diseases

Few pests bother cucurbits in the UK, although generalist pests, such as slugs and mice, can cause occasional problems. Pollen beetles (*Meligethes* spp.) are common, but not usually a cause for concern.

Powdery mildew (*Erysiphe cichoracearum*) is a common disease on outdoor squashes during the summer and autumn. Although unsightly, crops often grow through infections and the disease can be managed by cultivation of crop debris. Grey mould (*Botrytis cinerea*) is also common on developing fruit and is often aided by incidental damage to plants, either wind twist or hoe damage. Improving air circulation in the crop can help manage this disease. A number of viruses are also capable of attacking or stunting plants and are best avoided by clean seed, managing the aphid vectors and crop hygiene.

Leguminosae (Fabaceae)

Legumes are a well-known group of plants, due to their ability to fix nitrogen from the air with the aid of symbiotic bacteria in root nodules. As a consequence of this ability, many legumes are used for fertility building in organic rotations, either as part of ley mixtures (for example clover) or as stand-alone green-manure crops (such as vetch). Many legumes are, however, crops in their own right. Legume crops should be regarded as neutral in terms of removing nitrogen from the soil, as most of the fixed nitrogen will generally be removed in the harvested portion of a crop plant. In this respect they can be placed at any point in the rotation. However, they are best late in the rotation, so as not to waste nitrogen that might be available to more demanding crops earlier in the rotation.

Types and Varieties

Peas and beans are the two maincrop types that are generally grown to be harvested and eaten in vegetable-production systems. Beans fall into three main groups: field or broad beans, runner beans and French beans. The former two are better adapted to a temperate climate and the latter to warmer regions.

Peas: mangetout and sugar snap peas are popular for direct marketing, as they are rarely seen in supermarkets. They can be direct sown early in the season (or even over winter) and, thus, can be a useful early crop that can be grown under protection. Many varieties will need staking or can be grown up netting. They can be time consuming to pick over, but this is necessary to keep them producing. Vining peas are normally grown under contract for processing and selling as frozen.

Broad Beans: over-winter and summer types exist. Over-winter beans can produce a crop early in the season when not many other vegetables are available and can thus help from a marketing perspective. They are hardy and generally trouble free, but should be picked early and sold quickly, so as to avoid over-mature pods.

Runner Beans: are popular with gardeners and tend to crop prolifically, creating direct competition for commercial crops. They are generally cropped as an annual. In order to get the crop away early in the season they are usually planted in large modules and transplanted to the field when conditions are warmer. They are labour intensive and need to be trained up canes or other supports.

French Beans: both dwarf and climbing types of French beans exist. They prefer warmer conditions and will need to be raised in modules and planted out to gain an early start. The beans come in a range of sizes and colours and can provide an interesting vegetable for the direct-sales market. The dry beans can also be harvested for eating.

Soil Fertility and Rotation

Generally, peas and beans prefer well-drained and neutral soils. Apart from field beans, they will generally thrive best in warm sheltered sites.

They are best regarded as neutral in a rotation in terms of nitrogen demand but benefit from good levels of potash. They can be looked on as a break crop in a rotation.

Irrigation may be beneficial when pods are filling under dry conditions, especially for French and runner beans.

Weeds

Weed management can be difficult in peas and climbing beans and will generally be restricted to hand hoeing. In this case, preparation of the seedbed before sowing or transplanting to remove weeds is necessary, to avoid problems in the subsequent crop. Field beans can be more robust and can be comb-harrowed when small. Once established, beans will tend to shade out competitors.

Pests and Diseases

As a diverse range of crops, peas and beans naturally tend to suffer a diverse range of pests and diseases.

Beans do not generally suffer unduly from pest insects. Runner beans and French beans occasionally suffer from bean seed maggot (*Delia platura*), seen in the field as poor establishment, especially early in the season when conditions are cold and plant growth slow. This is generally avoided if plants are raised in modules. Similarly, slugs may be a problem in wet conditions and when trying to establish a crop from seed or just after transplanting. Black bean aphid (*Aphis fabae*) or other aphids can be a problem in some years, but normally predators will keep them under control. Pea and bean weevils (*Sitona lineatus*) characteristically eat the edges of leaves of field beans and peas, to give a serrated effect, but damage is rarely serious. Mice can cause nuisance damage by gnawing holes in pods.

Beans and peas tend to be susceptible to a range of root- and foot-rots, generally caused by fungal pathogens (*Aphanomyces*, *Phytophthora*, *Pythium*, *Fusarium*) that cause the plant to damp off or fall over, normally in cold, wet conditions when the plants are struggling to establish. Rotation will help to keep these diseases from building up in the soil. A range of foliar diseases may appear, including halo blight (*Pseudomonas phaseolicola*) and anthracnose (*Colletotrichum lindemuthianum*) in French beans and leaf- and pod-spot (*Aschochyta pisi* or *Mycosphaerella pinodes*) on peas. Many of these diseases are best managed by using clean disease-free seed and appropriate cultural measures, such as crop hygiene.

Broad beans suffer more than the other types with foliar diseases. Chocolate spot (*Botrytis fabae*) is worst on over-winter crops and in wet seasons, which can be mitigated by various cultural methods, such as wide row-spacing. Similarly, rust (*Uromyces fabae*) is common late in the season and can develop rapidly in warm weather, but has little effect on yield.

Solanaceae

The Solanaceae are one of the larger families of the plant kingdom and include many members that are used for a variety of purposes apart from eating (for instance tobacco). The family provides a range of crops, generally harvested for their fruits or tubers. The ubiquitous potato and tomato belong to this group, as do aubergines, chilli peppers and sweet peppers.

Types and Varieties

The various types of solanaceous plants cover a diverse range of crop plants. The more common ones are mentioned below. Many of them (apart from potatoes) require warm conditions to develop and produce fruit, and are best grown under protection.

Aubergines: form large, usually black fruits that are probably best for the direct-sales market. They require hot conditions to develop successfully.

Chilli Peppers: are best regarded as a specialist crop and will need to be grown undercover. The main potential is in direct sales.

Tomatoes: widely available and much in demand. In the UK they are best grown either in polytunnels or glasshouses, although some varieties can also be grown outdoors. In the latter case, wet cold weather is likely to ruin a crop before the tomatoes mature at the end of the season. Heated glasshouses can be used to considerably extend the season, but this practice can be regarded as questionable from an environmental perspective. A vast range of varieties exists and many are available as organic seed. Varieties and types can differ by fruit shape and size: beef (steak), plum, standard and cherry. However, there are also determinate and indeterminate plant types and some are trailing (but difficult to manage!). Many varieties are now F1 hybrids, although the more interesting traditional type varieties are open pollinated, allowing seed to be saved.

Potatoes: a diverse range of potato types have been developed: early, salad, processing, ware and bakers. Some are available or targeted at certain seasons, whilst ware potatoes are a staple all-year-round. A large number of varieties is generally available, many of them as organic seed. The various varieties represent a range of characteristics as to textures, flavours, colours, culinary characteristics, pest and disease resistance, and skin finish. It is, therefore, necessary to determine market requirements and outlets when choosing varieties.

Sweet Peppers: generally green in immature stages, they will ripen to a range of colours (green, yellow, red), which can be attractive in direct-marketing outlets. They need to be grown indoors.

Soil Fertility and Rotation

Solanaceous plants generally prefer slightly acid but well-drained soil. However, potatoes will grow in a wide range of soil conditions and types. In this case, harvest is difficult in heavy soils. A number of soil pests and diseases can seriously affect plant production in this family and at least a four-year rotation between solanaceous crops should be practised. A longer rotation is often beneficial.

Irrigation will be vital for protected crops and may aid outdoor tomatoes and help prevent scurf and growth cracks on potatoes.

Weeds

For the protected crops, adequate soil preparation will help to knock back weeds. Some hand weeding may be necessary from time to time at the base of plants, but this can be minimized with a targeted watering regime that doesn't allow weeds to germinate and establish. In some cases woven-plastic mulches are useful in controlling weeds (and help with moisture retention in the soil). Potatoes are generally grown on ridges, which are knocked down and built up with ridgers as a form of weed management. Once the crop canopy has closed, potatoes out-shade weeds and can be regarded as a cleansing crop in this respect. Perennial weeds, such as couch grass, occasionally become problematic in potato crops but should be tackled from a whole rotation point of view.

Pests and Diseases

There are a large number of pests and diseases of solanaceous crops and some are specific to each crop type. An overview of the most common problems is given here and specific pest or disease problems are discussed in greater detail under the specific crops in Chapter 12.

Many of the crops grown under protection suffer from a limited number of pests and diseases that can, in general, be managed by paying attention to cultural techniques, such as maintaining adequate water supply and ventilation. General rots (such as *Botrytis* grey mould) can sometimes build up as the season progresses or if conditions deteriorate under cover. Viruses can also be a problem on solanaceous plants, often transmitted by aphids or by poor hygiene.

Potatoes are an important cash crop for many organic growers and, from this perspective, are the most challenging as regards pest and disease attack. Insect pests are usually incidental in this crop, but aphids can be important vectors of virus diseases. Slugs can also be a problem, especially on heavy soils where they attack the tubers. Of the diseases, late blight (*Phytophthora infestans*) is the most challenging disease, frequently observed on potatoes and outdoor tomatoes. On potatoes, skin-blemish diseases, like black scurf (*Rhizoctonia solani*) and common scab (*Streptomyces scabies*), are becoming increasingly important and are best controlled by rotation or cultural means.

Umbelliferae (Apiaceae)

Umbellifers are well represented in the UK flora as annual, biennial or perennial herbs, many of which are common on roadside verges. Many are also known for their medicinal properties or essential oils. A few have been adapted as vegetables, using many parts of the plant: roots (carrot), stem (celery), leaves (parsley).

Types and Varieties

There is a range of umbellifers in common cultivation and the most commonly grown are mentioned below.

Celery/Celeriac: generally a demanding crop to grow organically due to nutrient dynamics. Good irrigation and/or water-retaining soils are an advantage when growing this crop. Leaf spot (*Septoria apiicola*) is a common seed-borne disease in this crop and it can be difficult to obtain clean organic seed. Slug attack can cause problems, but only a limited amount of plant resistance is currently available for either of these problems. Celeriac is the same species, but the root is used rather than the stem.

Carrot: varieties have been trialled under organic conditions in the UK over the past decade or so. Although grown in diverse systems, successful varieties usually share a need for good early vigour to emerge quickly and out-compete weeds. Varieties can be grouped according to maturity: first early (sown in October under cover for harvest in May), second early (sown in January or February for harvest in June–July), early maincrop (sown May for harvest in September–November) and maincrop (sown May for harvest November onwards). Second earlies and early maincrop are often sown in organic systems to avoid carrot fly, in which case Nantes varieties are commonly used. Other types include Chantenay, Berlicum and Autumn King types. Nantes types are more susceptible to *Alternaria* (leaf spot) and Chantenay more resistant. Autumn King varieties tend to be more resistant to motley dwarf virus. Nantes can be grown for prepacking and fresh market, Chantenay for fresh market and processing, while Autumn King are mainly used for the fresh market. Various pieces of specialist equipment exist for sowing, weeding and harvesting this popular crop.

Parsnip: is a native, winter vegetable that can be stored in the field and harvested over a long period. Parsnips need to be resistant to canker; powdery mildew can occasionally be a problem. Carrot fly can also cause serious damage, but market specifications are often more lenient than with carrot. They are generally very slow to germinate, so need good predrilling and pre-emergence weed control.

Soil Fertility and Rotation

Umbellifers are generally not too nutrient-demanding and are often placed after brassicas in a rotation. Where soil-borne diseases have been a problem, a long rotation of five years or so is best to help manage the diseases. The root crops do not generally do well on heavy soils and are easier to manage on light soils. Large numbers of stones might also cause fanging in carrots and parsnips. They prefer a neutral pH.

Irrigation might be needed to ensure even germination of carrots and parsnips and is beneficial to celery and celeriac.

Weeds

Umbellifers are often slow to establish and this can cause problems for early weed management. Stale seedbeds are often used and, in the case of carrots, flame-weeding just before emergence. If weeds are not suppressed at the initial stages, brush weeders or steerage hoes can be used shortly after emergence and in the first month. Hand weeding is also effective, but time consuming and expensive, especially as the crop matures. Large amounts of weed can hamper manual harvesting or mechanical harvesting machinery if it is not removed.

Pests and Diseases

Carrot fly (*Psila rosae*) is the principal insect pest of organic carrots, the maggots rendering carrot roots unsellable and also causing problems in celeriac and parsnips. Forecasting services are available that predict when carrot fly adults are likely to be laying eggs and these can be used to time the placing of crop covers or to determine the best sowing times to avoid peak fly activity. Aphids (*Cavariella aegopodii*) are occasional pests that might cause damage in some years to carrots and transmit the motley dwarf virus, which affects carrots and parsnips. Celery fly (*Eulieia heraclei*) causes unsightly blotch mines on leaves of parsnip and celery, but does not generally cause damage. Slugs can occasionally cause problems in the various crop types.

Foliar diseases of umbellifers include leaf blight (*Alternaria dauci*) on carrots, leaf spot (*Septoria apiicola*) on celery and powdery mildew (*Erysiphe heraclei*) on carrots and parsnips. The former are best managed by using clean, uninfected seed. Canker (*Itersonilia pasinaceae*) affects parsnip roots, violet root rot (*Helicobasidium purpureum*) affects carrot and parsnip roots, whilst cavity spot (*Pythium violae, Pythium sulcatum*) affects carrots. These diseases are best managed by cultural means, such as maintaining a long rotation between crops and harvesting the crops early before the disease develops. Celery is also susceptible to pink rot (*Sclerotinia sclerotium*).

VARIETY TRIALLING

Given the large range of vegetable varieties available and the range of desirable traits, vegetable variety-trialling provides potentially valuable information that allows growers to select varieties suitable for their production needs and market outlets. Many new and old vegetable varieties are offered for sale in each season and, consequently, growers need advice on varietal characteristics and performance in order to choose the most suitable ones. Before choosing, growers should decide which varietal characteristics are appropriate for their marketing strategy. They should also identify what pest, disease or growing problems might be encountered within their particular location and production system. They then need to match these requirements with information on varieties, in order to choose the ones that will enable them to meet customer specifications, provide the necessary quality characteristics and be resistant to troublesome pests or diseases.

Currently, the main source of varietal information is through seed catalogues, based on information from seed-company testing programmes, and through various national testing schemes, normally based on information obtained through trials carried out under the auspices of NIAB (formerly National Institute of Agricultural Botany) or agricultural colleges. Variety-testing trials are normally carried out on research stations or on larger commercial farms using replicated field trials that provide statistically analysable information on performance. They are often carried out at different sites or in different seasons to get a measure of the reliability of the data and of the adaptability of the variety. General dissemination of trial results is through the trade press, in booklets, in fact-sheets, at open days and by communication between growers. NIAB also publishes an organic vegetable variety handbook, which details the results of variety trials in commercial situations (*see* Further Reading) and this information is also available through the Centre for Organic Seed Information (COSI). Accurate internet-based information is becoming increasingly and freely available.

Evaluating celery varieties on a commercial farm

This system plays a valuable part in helping growers make varietal choices, in both conventional and organic growing systems. However, organic farming systems arguably represent a more diverse range of growing environments, which might not be adequately represented by sites in the current trials programmes. For instance, varieties adapted to labour-intensive, diverse planting regimes, typically selling into farm shops or box schemes, might not be the same as those suitable for large-scale production, typically sold through supermarkets. Organic growers also often value varieties for a range of characteristics that are not so important in conventional systems and are often difficult to quantify in large field-scale trials. For example, varietal vigour is often important to organic growers where nitrogen (N) availability is limited. A good root system to access available nutrients can therefore be important, as can a dense, closed canopy with good weed-smothering abilities. Pest and/or disease resistance is often important in organic systems, where no preventative or curative sprays are permissible or available. Taste and other similar qualities are also very important to organic consumers and these might not be adequately assessed in large field-scale trials.

For all these reasons, organic growers should also seek to test varieties in their own growing systems, in addition to using the results from the more conventional testing programmes. This can best be done by planting small areas or plots to new varieties and recording varietal performance by taking notes on important characteristics. Any promising varieties should be tested in two or more seasons to ensure the results are reliable over a number of seasons. In this latter case, note taking is essential, in order to compare performance over a number of years. Finding out from other organic growers how varieties have performed can also be important and sharing information could help to speed up the process of selecting varieties suited to particular areas.

ORGANIC SEED- AND PLANT-RAISING

Establishment of Vegetable Crops

Varieties with good seedling characteristics are better in organic systems (*see* above). Positive characteristics include resistance or tolerance of seed-borne diseases, high germination rates and high seedling vigour. Apart from the variety characteristics, vegetable crops will normally be obtained as seed or transplants and it is vital that these are healthy and vigorous, in order that the crop be given the best possible chance of establishing well.

Vegetable crops are generally established either by direct sowing or by transplanting them into the final growing position. The latter technique is usually more expensive and can require outlay in terms of equipment if raised on the farm. This is especially true where plants are grown in protective structures, such as polytunnels or glasshouses. Before transplanting, plants may need to be 'hardened off' for a period to acclimatize them to field conditions and, in many cases, will need to be watered in, especially under dry conditions. The basic techniques are described briefly below, together with an overview of the advantages and disadvantages of each method.

Direct Sowing

Some crops, such as carrots, are always directly sown into their growing position. Others, such as onions, can be direct sown or raised as transplants, depending on the facilities and farm system.

Bare Root Transplants

These are either grown in a nursery bed and then lifted and planted into their final position in the field or they can be raised as rows of seedlings in trays, for example cabbage and leeks. The latter method requires more equipment and consumables. Bare-root transplants are not so popular now, because lifting and replanting can lead to root damage and irregular establishment, unless a great deal of care is taken during the transplanting operation. The transplanting operation can itself be more complicated, as it is necessary to stop the bare roots drying out during the process.

Module Transplants

These are grown in modular trays or blocks of growing media or compost and the whole plug or block is placed in the final planting position. Seeds are generally sown individually in their own block or module, so that the root system of each plant is separate and, consequently, undamaged at transplanting. Modules can also be multiseeded (for example onions or beetroot) to reduce costs. In some circumstances modules can be 'held' until conditions are optimum for planting the crop.

When growing seedlings in modules or trays, there are a large number of management factors to take into consideration, including:

Containers: transplants are generally produced in multicelled trays, which are available with a range of cell sizes. The best cell size to use will depend on the crop being grown and the length of time required. Organic plants generally perform better in larger cells that hold more nutrients and moisture.

Media: plant quality will ultimately depend on the suitability of the media in which they are grown. The media mix should contain adequate nutrition for the plant during raising and retain sufficient moisture to allow for seedling development. The media should, ideally, be peat free, although there are organically certified composts that use non-mined peat (for example that has been reclaimed or filtered out of drinking water). There is a range of non-peat alternatives available based on

Self-propelled leek planter

agricultural by-products (such as coir) or composted green-waste. These media are less water-retentive, will therefore require more frequent watering and may not hold together as firmly as peat-based media. This will require more care in transplanting.

Plant Feeds: during production, plants might need feeding to promote development. Such fertilizers would typically provide major or minor nutrients in a more or less readily available form and should, in any case, be limited to those permitted under organic standards. They are usually applied as liquid or foliar feeds and nowadays are often plant- or seaweed-based, as opposed to animal by-products (say blood and bone meal) as was formerly the case.

Watering: thorough and even watering is necessary and should be done to just before drip point to avoid leaching nutrients. Overwatering and waterlogging is a common problem and should be avoided, as it causes poor growth and can encourage damping-off diseases. Holding plants off the ground with good air circulation can help in this respect. Corners and edges dry out more quickly and need particular attention. Older plants will require more water and may have to be watered more than once per day.

Light and Heat: both need to be adequate, but not excessive for good plant growth. Shading might be necessary on warm sunny days and excessive heat can quickly build up in glasshouses or polytunnels unless adequate ventilation is supplied. If excessive plant elongation ('leggy plants') is a problem, brushing plants might help, as can reducing temperatures.

Pests and Diseases: minor pests such as sciarid flies, white flies and damping-off diseases can occasionally be a problem in plant raising, but are generally avoided by rigorous hygiene and an appropriate watering regime. If problems do occur, there is a range of biological controls that can be applied and that usually work well, especially in enclosed structures and if given time to establish.

Advantages and Disadvantages of Techniques

Transplanting is generally preferred over direct sowing in organic vegetable production, where it is an option, because it produces better and more trouble-free crops and potentially utilizes resources more efficiently. When using transplants, the crop occupies the field for a shorter period and it will, therefore, generally require less weed control if the land is well prepared. As crops are present for a shorter period, pests – and especially diseases – are often reduced in severity, because plants are more developed when they are transplanted and there is less time or opportunity for epidemics to become established and develop. Transplanting also potentially maximizes cropping in the exploitative phase of the rotation and allows the direct establishment of target-crop populations, which can be important to meet size specifications or programmed maturity dates. Indeed, transplanting is often necessary to achieve early maturity times. Additionally, less seed is generally needed and this might be important, as organic seed typically costs more than comparable conventional seeds.

Only smaller vegetable production units usually use bare-root transplants. This is because they generally allow more flexibility for these types of vegetable-growing systems. In terms of expense, there are no transport costs (which can be proportionally large on small orders of bought-in module transplants) and, at this scale, there can be some savings in raising material if done in a nursery bed. In addition to these advantages, bare-root transplants generally allow greater flexibility in variety choice (especially of small quantities of unusual varieties), in planting date (as nursery conditions can be tailored to deliver plants to the main holding at the right time) and can be more hygienic (no risk of introducing soil-borne disease such as clubroot). However, to offset against this are possible weeding costs, occasional problems with damping-off diseases (such as wire stem (*Rhizoctonia* spp.)) and potentially more complicated and slower planting regimes.

Modules are generally used in larger-scale field production units. At this scale costs can be reduced by buying in large quantities of transplants from specialist plant-raising nurseries. Specialist and automatic machinery also exists for planting modules and this system of plant raising generally lends itself better to automated or semi-automated planting systems, thereby reducing labour costs.

Legal Considerations

Organic Seed

The European Council Regulation on organic agriculture requires the use of organic seed (and propagating material) where it is available. This regulation (EEC No. 2092/91) allowed for the use of untreated conventional (non-organic) seed until 31 December 2003 if a grower could demonstrate to their organic certification body that they were unable to obtain suitable organic seeds. Subsequent to this, Commission Regulation (EC) No. 1452/2003 has maintained the derogation beyond the date of 31 December 2003, because there would not have been sufficient amounts of organically produced seed for certain crops. Therefore, if organic seed of a particular variety is not available, growers may use untreated non-organic seed, provided they have prior permission from their organic certification body and no reasonable equivalent organic variety exists.

To improve the transparency of this system, the Commission Regulation required each Member State to set up a database showing the availability of organic seed. In the UK, DEFRA considered that a website offered the best way to make up-to-date information on organic seeds widely available to growers within the UK and this requirement is met by the OrganicXSeeds online database. In addition to this, The Centre of Organic Seed Information (COSI) has also been set up, with funding from DEFRA, to promote the dissemination of information on organic seed within the organic industry. COSI also provides information on UK variety performance data for different crops (based on variety trials data) and information on regulations affecting the use of seed in organic farming systems. Organic farmers now have access to both websites, which provide them with information on the availability of organic seeds within the UK and some information on variety performance. People without internet access can either call their certification body or the Soil Association (SA), as manager of the database, to request a search of the database to be undertaken on their behalf.

Seed Derogations Procedure

Growers must look at the internet database to check whether the variety of seed they want to use is available organically. They must also check that an equivalent substitute variety is not available either. The information to do this is available through OrganicXSeeds and COSI. If no suitable variety is available, they can apply to their certification body for a derogation to use non-organic seed (this can be done online from the websites). If an alternative variety is available they must buy at least a percentage of that variety and are encouraged to suggest that the supplier state an interest to the seed merchant in the other variety being produced organically.

Plant Raising

Growers must use permitted growing media and any additional liquid feeds must also be either permitted or a derogation obtained before their use if they are restricted. Some inputs are prohibited under organic standards and cannot be used under any circumstances. Bought-in transplants must have been raised on a registered organic unit in permitted growing media.

Chapter 4

Fertility Building

Soil is a key factor in organic systems and this is especially true of vegetable-production systems, as these make particularly severe demands on the soil. In this chapter we aim to explain the underlying concepts of soil fertility and outline techniques for enhancing and maintaining soil fertility in vegetable-production systems. The use of composts and manures will also be discussed, as will techniques for using cover crops, both to prevent leaching and to provide nitrogen by fixation.

WHAT IS SOIL FERTILITY?

In an agricultural context, a fertile soil can be defined simply as one that supplies sufficient nutrients to the crop to allow maximum growth and yield. However, this is a very narrow definition and does not encompass the whole range of soil functions. For instance, an intensively managed soil may continue to provide sufficient nutrients to a crop through additions of synthetic fertilizers, even as the physical condition declines. This decline might be seen as reduced workability, increased erosion and nutrient loss, as well as declining biological activity. A degraded soil showing such characteristics could hardly be considered fertile or maintain sustainable production over the medium- to long-term. A broader definition of soil fertility – more compatible with organic principles – would encompass all soil functions; to include biological activity and physical condition, as well as the ability to provide nutrients to the crop. This interpretation of fertility can also be associated with the idea of soil health. A healthy soil will perform all of the above functions, while an unhealthy or sick soil may still supply nutrients to the crop but would fail to meet the broader ecosystem considerations. Although it is useful to consider soil fertility in this wider context, the complexity of soils means that it is necessary to discuss the different aspects of soil fertility separately.

Chemical Fertility

The chemical fertility of the soil is probably the most easily understood. Plants require a range of nutrient elements in various quantities, which they obtain from the soil. These elements can be measured in soils and the amount related to crop growth. Where a deficiency occurs, fertilizers can be added to rectify the problem. Although this story appears deceptively straightforward, it is somewhat of an oversimplification and, if taken too literally, can lead to longer-term soil problems. For this reason, we have developed a more complete picture of chemical fertility below.

Plant nutrients are found in a number of forms in the soil. These are:

- as primary minerals originating from the bedrock or soil parent material
- as secondary minerals formed by chemical weathering
- in organic matter
- as ions, either in the soil water or attached to organic matter and clay minerals by electrostatic attraction.

Nutrients are also present as constituent parts of living soil organisms. The strength with which different elements are held by different soil components varies, making it very difficult to assess how much of a nutrient in a soil will be available to plants at a given time. Nutrients held in primary and secondary minerals and some humus may be only available over time scales of tens or even hundreds of years, while those held electrostatically on the surfaces of clay minerals and organic matter are quickly available. Soil micro-organisms hold some of the most readily available nutrients; their populations turn over rapidly and, as they die, the nutrients they contain can be released into the soil water.

All this means that the total content of nutrients in the soil, especially of macronutrients – nitrogen (N), phosphorous (P) and potassium (K) – is of little use in determining crop nutrition. An extreme example is provided by the 'Lower Cook laterites' in Australia. Though very high in total phosphorus (0.5–2 per cent), this is mostly in forms not available to plants and these soils are classed as phosphorus deficient. Even with elements for which total content is a useful measure, for instance copper (Cu), the situation is not straight forward, as, in this case, high molybdenum (Mb) levels can reduce the availability of copper to plants, although this can in turn be alleviated by the use of sulphur-containing (S) fertilizers. Despite the problems of quantifying the amounts of nutrients available to plants, successful fertilizer recommendations have been developed for conventional systems. These have allowed continually increasing yields, though not always without environmental consequences.

An organic farmer needs to understand the chemical fertility of the soil as much as a conventional farmer. Though the management options are slightly different, with a more holistic approach aimed at managing the whole soil system, the aim is similar. That is, to maintain concentrations of soil nutrients at a sufficiently high level to maintain crop yields without accumulating nutrients to the extent where losses to the environment can cause pollution problems.

Biological Fertility

The importance of soil organisms in the functioning of soils, both in cycling nutrients and maintaining soil structure, has been recognized since the nineteenth century. However, the study of soil organisms, with the exceptions of some pests, diseases and earthworms, has been a neglected area of investigation. There are two reasons for this. Conventional crop yields have been successfully increased over the last fifty years and continue to increase, simply by managing 'bagged nutrients'. The second reason is the difficulty in studying soil micro-organisms. Though something of the functions of soil micro-organisms are known, it is only recently, with the development of molecular techniques, that it has been possible to even get a reliable estimate of some basic facts, such as the number of species in soil.

Bacteria are the largest microbial group in agricultural soils, in terms of populations and biomass. Arguably they are also the most important group. They are responsible for the breakdown of organic matter and the release of nutrients to the crop. They also play a vital role in maintaining soil structure, through the production of extracellular polysaccharide gums, which stick soil particles together.

The second largest group are the fungi. In drier, more acid and particularly in undisturbed soils they can be more important than the bacteria. Like the bacteria, they are responsible for the breakdown of organic matter and the release of nutrients. Fungi are able to break down resistant organic matter that bacteria cannot. As with bacteria, they are important in maintaining soil structure, but their mode of action is slightly different. Fine strands of fungal hyphae enmesh soil-aggregates, holding them together like a string bag. These fungal hyphae are more long-lived than the bacterial polysaccharides and form an important part in maintaining soil structure. A sub-group of the fungi, called mycorrhizal fungi, are also important for their role in forming a symbiotic relationship with the roots of many crop species. This relationship assists the uptake of nutrients, particularly phosphorus, and gives some protection from pests and disease. However, in vegetable-production systems in particular, their importance is still uncertain. Some crops, such as the alliums, are strongly dependent on mycorrhizal fungi, while others, such as the brassicas, do not appear to form mycorrhizal associations. It is also generally observed that intensive cultivations and high nutrient levels, conditions often found in vegetable production, suppress mycorrhizae. Some soil fungi are also responsible for some troublesome plant diseases, including *Verticillium*, *Pythium* and *Rhizoctonia*.

As well as the bacteria and fungi, other groups of soil micro-organisms are important in terms of soil functions and crop management. The actinomycetes are a group with some fungal and some bacterial properties. Their importance varies from soil to soil. They are uncommon in soil that is frequently waterlogged, but are resistant to drought and can dominate soils after prolonged drought. They also favour alkaline soils and are few in number below pH 5. They perform a role similar to the bacteria and fungi, breaking down organic matter, with a particular importance in the breakdown of chitin, the main component of insect exoskeletons and fungal hyphae. However, there are also some which cause important diseases, such as the causal agent of potato scab (*Streptomyces scabies*). Incidentally, they are also responsible for the characteristic earthy smell of moist soil.

Other groups of soil micro-organisms include the protozoa, algae and viruses. The protozoa (mostly amoeba), though small in terms of biomass when compared with the bacteria and fungi, are important because most types feed on soil bacteria, forming a vital link in nutrient-cycling within the soil. In some laboratory experiments, grazing by protozoa has been shown to increase nitrogen mineralization and nitrogen uptake by plants by as much as 75 per cent. Some protozoa also feed on fungal hyphae and spores, including those of diseases like take-all (*Gaeumannomyces graminis*) and root rots (such as *Cochliobolus sativus*). Algae have a minor role in soils, while the importance of others, such as viruses, is unclear, though some cereal and vegetable viruses, such as lettuce big-vein virus and some barley mosaic viruses, are soil borne and very difficult to control.

Soils contain a large number of macro-organisms, which are important in soil functioning. Nematodes are small worm-like animals, 0.5–1.5mm long, that perform a wide range of roles. Some nematodes are predatory, forming an important link in the nutrient cycles of soil, while others feed on organic matter, helping breakdown and nutrient release. However, some are responsible for the most intractable of pest problems. These include potato-root eelworm and root-knot eelworm, which can produce resistant cysts that may persist in the soil for years, making control by rotation very difficult. In the case of organic farming, root-knot nematodes are a particular problem in red clover and have resulted in the abandonment of red clover in rotations in some locations.

Larger animals in the soil are mainly arthropods. They have an important role in the initial stages of organic-matter breakdown, mainly through dividing it into smaller particles, giving a larger surface area for fungi and bacteria to act on. Where they are excluded, organic-matter breakdown proceeds at a much slower rate. There are some species that are important agricultural pests, such as the wireworms and leatherjackets. The group of soil organisms most obvious to the casual observer are the earthworms. They form an important role in the initial stages of organic-matter breakdown, through dragging surface material into the soil profile and dividing it into smaller particles. They also form deep channels, which are important for drainage, particularly in minimally cultivated soil.

All of these groups of organisms interact in complex ways by forming intricate networks or foodwebs. Indeed, the soil has been termed the poor man's tropical rain-forest, because of the diversity of the organisms that live there and the complexity of their interactions. How much of this diversity is required to maintain soil functions is not clear, though recent research suggests that overall diversity is less important than functional diversity. Most agricultural activities, such as cultivations, liming and use of fertilizers, will reduce diversity. Though the systems are generally very resilient and will recover from most disturbances given a moderate amount of time, some chronic disturbances, such as heavy metal contamination, can cause long-term damage to soil biology.

Physical Fertility

The physical state of the soil is determined largely by its structure. Soil structure is the arrangement of solids and air spaces in a soil. It is determined to a large extent by the soil texture. That is, the proportions of sand, silt and clay particles determine the soil type. This is usually illustrated using a soil texture triangle (*see* Fig. 4.1).

Soil texture can be assessed accurately by laboratory methods, but a reasonable assessment of soil texture can be made using a field technique described in the section on assessing soils below. Texture is important, because the different size particles have different properties. For instance, sand does not cohere and so a sandy soil will be vulnerable to erosion by wind or water. Clay, on the other hand, binds strongly to itself, giving a robust soil that is resistant to erosion, but which can form hard clods and layers that roots struggle to penetrate.

Sand, silt and clay combine with organic matter into larger clusters known as aggregates. These aggregates combine further to form the crumbs or clods in the soil. Their size, shape and distribution determine the soil structure. As well as affecting root growth, soil structure is important for allowing the movement of air and water, which has effects on nutrient availability, workability, stock-carrying capacity and pest and disease severity.

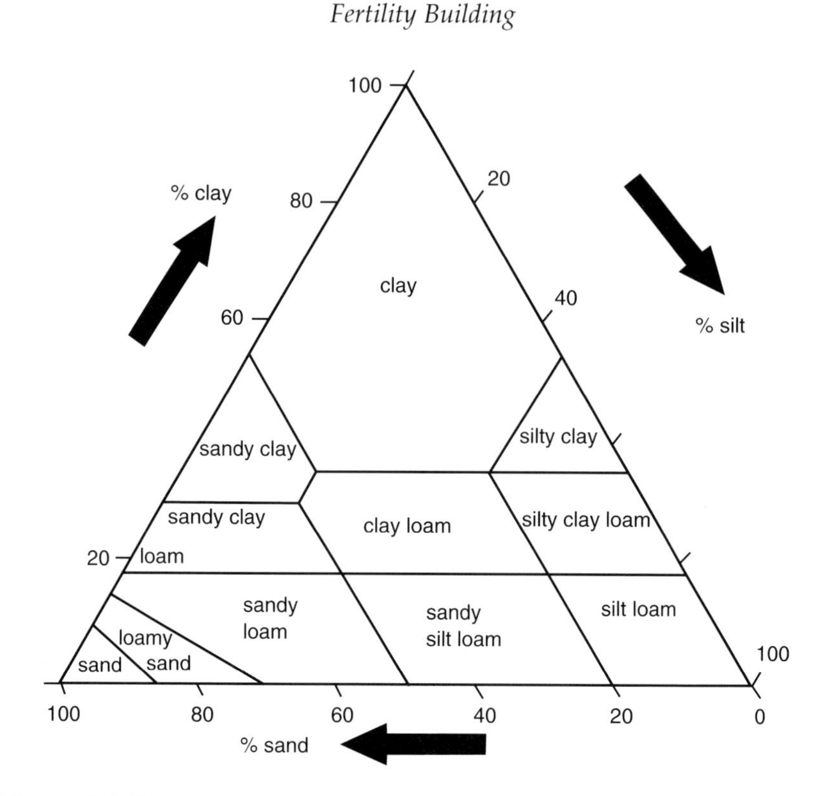

Figure 4.1 Soil texture triangle as generally used in the UK

A poorly structured soil may have a number of problems, including:

- in wet conditions it will be slow to drain and, thus, will warm more slowly in spring
- water availability will be reduced in drought conditions because of limited root penetration
- surface capping of sands and silts can reduce or even prevent emergence of seeds
- surface capping reduces rainfall infiltration and thus can result in severe erosion of vulnerable sand and silt soils
- reduced availability of air and reduced temperatures will slow microbial activity, which will limit nitrogen mineralization
- limited root growth will reduce access to nutrients
- slow crop growth can increase susceptibility to certain pests and diseases
- increased vulnerability to poaching, perhaps leading to animal foot-problems

- extra cultivations will be required to correct damage, while more power will be required to work compacted soil and working windows will be reduced, as soil is slower to dry and more vulnerable to damage.

Soil structure is particularly important in organic systems, as problems associated with poor structure cannot be compensated for by increased fertilizer use or the use of biocides. It is also no longer economic to alter soil texture by altering the proportions of sand, silt and clay, though in the past the practice of marling light-soils was widespread.

MANAGING SOIL FERTILITY

Management of soil fertility in conventional systems has been based on management of soil chemistry through the use of synthetic, soluble fertilizers and lime, with some limited attention paid to soil structure. Crop residues and animal manures have often been seen as a waste problem, rather than a valuable input. In contrast, soil management in organic agriculture has been based on the use of these organic materials that the conventional approach has regarded as problems. More emphasis is also placed on managing a soil's physical structure and on promoting biological activity, the importance of which has been outlined above.

Although organic and conventional systems are very different in their emphasis, they both have the same ultimate purpose, namely the provision of plant-available nutrients in sufficient quantities to produce vigorous crop growth. The more holistic approach to soil management in organic systems and the greater reliance on biological activity to supply nutrients to crops, which is itself dependent on good soil structure, means that is it unrealistic to separate chemical, biological and physical management. Instead, the different management practices used by organic farmers to build and maintain soil fertility will be examined in turn. Soil testing and nutrient budgeting will also be briefly discussed.

Building Soil Fertility

Rotations

The use of rotations is a fundamental part of organic farming. Apart from their importance in managing soil fertility, they are also vital for the control of pests, disease and weeds. Organic rotations consist of fertility-building phases, when legumes are used to increase soil nitrogen, and fertility-depleting phases, when cash crops utilize the accumulated nitrogen. In many mixed arable systems the two phases are distinct and each may last several years. In stockless rotations the fertility-building phases are shorter, as they produce no direct economic return and there may be more reliance on imported nutrients. In intensive vegetable-rotations there is often a breakdown in the clear distinction between a fertility-building and a fertility-depleting phase. The wider range of crops grown gives more flexibility in rotations and fertility building is often done with short-term leys and green manures being grown for as little as a few months, fitted in between cash crops.

Generally speaking, crops with a high nutrient demand, such as potatoes and cabbage, are placed immediately after a fertility-building period, with less nutrient-demanding crops, such as carrots and onions, placed later in the fertility-depleting phase. Careful selection of crops, combined with the effective use of cover crops in winter, allows the maximum exploitation of soil nitrogen, while supplementary nutrients, in the form of manures, composts and fertilizers, can be added at any point in the rotation. Fig. 4.2 shows an intensive vegetable rotation, with soil-fertility points highlighted.

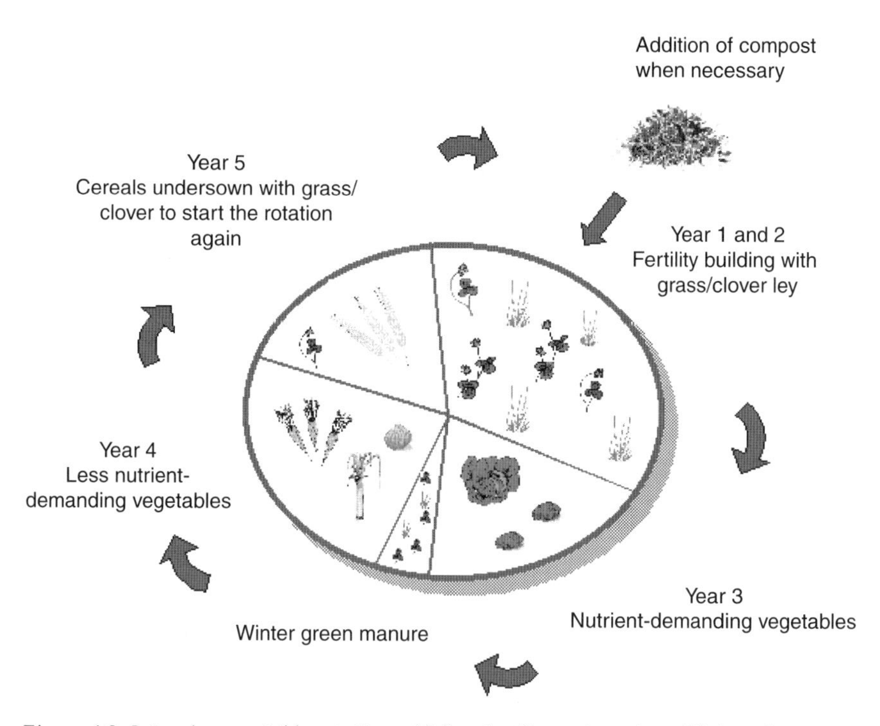

Figure 4.2 Intensive vegetable rotation with key fertility points: 1 – addition of compost; 2 – short fertility-building period; 3 – undersowing to establish ley; 4 – long-term ley

As organic growing is based on a whole-systems approach it is important to plan the rotation in advance, so that nutrient budgets, marketing, weed control and so on can be integrated. However, there must also be flexibility in rotations to allow for unexpected weather, changes in soil fertility or alteration to the market conditions.

Rooting Depth

Organic standards require the growing of crops with different rooting depths, in order to fully utilize the soil resource. This is particularly important for nitrogen, as, especially in lighter soils, it can be leached below the rooting depth of some crops. Deep-rooted crops should be grown after fallow periods, when leaching will be high, while shallow-rooted crops can be grown after a ley or green manure, which will concentrate nutrients in the upper layers of the soil. In the example shown in Fig. 4.2, if it was too late to establish a vetch after a harvest of potatoes, significant leaching could occur. However, some of this leached nitrogen could be recovered if the vetch green-manure following the potatoes was replaced with deeper-rooting cereal rye. Unfortunately, there is no absolute consensus on the rooting depth of crops, making such decisions less than straightforward. Table 4.1 gives an indication of the rooting depth of some common crops. As well as the crop type, rooting depth depends on soil depth, growing conditions and the length of time a crop is in the ground. Thus, an early carrot crop will not root as deeply as an autumn crop, while a cover crop, which is grown for four months, will not root as deeply as one grown for six. In addition, on a shallow or very freely draining soil the leached nitrogen may go below the maximum rooting depth of any crop before it is possible to recover it.

Table 4.1 Root depth of some common crops

Shallow	Medium	Deep
Cauliflower	Broccoli	Beetroot
Celery	Brussels sprouts	Cabbage (white)
Lettuce	Cabbage (other)	Fodder rape
Leeks	Carrots	Parsnips
Onions	Potatoes	Red clover
Peas	Spring cereals	Winter cereals
Spinach	Squash	
	Sweetcorn	
	Vetch	
	White clover	

Soil Structure

Another soil fertility factor that should be considered when designing a rotation is soil structure. Severe soil compaction can result in large yield-reductions in vegetable crops, exceeding 50 per cent in some cases. If soil structure has been severely affected, by harvesting in poor soil conditions for instance, it may be worth considering introducing a green manure, such as cereal rye, to help improve structure before planting the next cash crop, even if this was not in the planned rotation. Some crops, such as sweetcorn, may be less sensitive to poor soil structure but there is little available information to indicate which are the most vulnerable crops.

Intercropping

Though it is usual to separate the fertility-building and -depleting phases, there have been attempts to combine the two, avoiding the requirement for a period of zero economic return during fertility building (especially in stockless vegetable systems). The most common method is under-sowing a cereal with a ley crop. This means the ley is established much sooner and, therefore, needs to be in place for a shorter time or, alternatively, fixes more nitrogen in the allotted time. However, establishment can be poor, especially in dry years, and the ley can suffer from compaction caused during harvesting of the cereal.

Compaction in wheelings between beds in onions

There are some well-known intercrop systems, such as mixed beans and maize, but these are few in number and often suited to farming systems with plenty of labour available. In vegetable systems, a number of experimental systems have been tried, where the crop is either planted into a fertility-building crop or a fertility-building crop is established between the vegetable rows. They have met with varying degrees of success. One approach is to have semi-permanent fertility-building crops into which vegetables are planted; the problem here is reducing competition from the fertility-building crop. Where the fertility-building crop is killed at planting this is not a problem. Establishing a fertility-building crop between rows after planting vegetables is similar to the under-sown cereal method. However, with all three methods there is usually a reduction in yield of the vegetable crop, which increases as the time of overlap between the fertility building and vegetable parts increases, although the degree of yield reduction varies depending on the vegetable. Whether the yield reduction is compensated by improved fertility building or other associated benefits requires whole-system research, which has not been carried out.

Legumes

Legumes are the cornerstone of fertility building in organic farming systems. They are the main, and in some cases only, way of adding nitrogen to the system, apart from small amounts deposited from the atmosphere. Though several tonnes of nitrogen per hectare are stored in agricultural soils as organic matter, most of this is in the form of humus, which is only broken down over time-scales of years and decades. Without regular additions of nitrogen, crop yields rapidly decline to low levels. Alternatives to legumes, such as importing animal manures from conventional or other organic systems or the use of green-waste compost, can be questioned from a sustainability, health or ethical standpoint and are often economically unviable.

Legumes are usually incorporated into the rotation in distinct fertility-building periods. Many organic systems are mixed, supporting both livestock and cropping. In these cases legumes are incorporated into longer ley periods, which perform the dual roles of fertility building and fodder production. The most common legume for this purpose is clover, either white (*Trifolium repens*), more common for grazing, or red (*T. pratense*), more common for conservation in silage. Sometimes a mixture is used. They are usually grown with one or more grass species when used for grazing, though pure red clover leys are common for silage production and especially for cutting and mulching in stockless systems. In the drier parts of the UK, and more commonly in Europe, lucerne (*Medicago sativa*) is sometimes used. It is highly productive, but slow to establish and not suitable for grazing.

Where animals are not included in the farming system or they are only few in number, there is no demand for fodder production. This is common in vegetable-production systems, especially in the more specialized large-scale field vegetable-production systems. In this situation, leys do not produce any direct economic benefit and so they tend to be of shorter duration, perhaps a year or less. In these cases, leys need to establish quickly and often other species, such as vetch (*Vicia sativa*), are used. Such 'green manure' crops are discussed below. Grain legumes, such as peas and beans, are important parts of many organic arable rotations and may be important in some vegetable systems. Though they fix some nitrogen they are not normally regarded as fertility-building crops, as much of the nitrogen they fix is removed in the grain, so they make only a small contribution to soil fertility. They may even deplete soil nitrogen in some cases.

The amount of nitrogen fixed from the atmosphere by both fodder and grain legumes is difficult to quantify accurately. There is a reasonable relationship between yield and nitrogen fixed, but it is different for different species. This means that a legume grown between April and September will fix much more nitrogen than one grown between October and March. Clovers tend to obtain the greatest proportion of nitrogen from fixation: white clover, for instance, usually obtains 70–99 per cent of its nitrogen from the atmosphere. In contrast, grain legumes, particularly those of the *Phaseolus* group, obtain less than half of their nitrogen supply from fixation, the rest coming from soil nitrogen. Table 4.2 gives ranges of fixation by some common legumes.

Fixing nitrogen does, however, incur a cost. Up to 20 per cent of carbon fixed by the plant may be supplied to the rhizobium bacteria, which actually perform the fixation. Because of this, legumes preferentially use soil nitrogen if it is available, although most are poor competitors for nitrogen. As a result, anything that adds nitrogen to the soil will reduce fixation, while removing nitrogen will encourage it. Table 4.3 shows the effect of common management practices on N fixation.

Nitrogen fixation in leys declines with age, as nitrogen accumulates in the soil and the proportion of clover declines. Table 4.4 shows the effect this has on fixation and underlines the value of short leys as compared with longer leys.

Interestingly, research has shown that under elevated levels of atmospheric carbon dioxide, legumes, including pasture legumes, both increase nitrogen fixation and, in some cases, increase the proportion of nitrogen obtained from the atmosphere, both when grown alone and together with non-legumes. This should lead to increasing productivity of organic leys in the future, as atmospheric carbon dioxide concentrations continue to increase, providing soil nutrients and water do not become limiting.

Table 4.2 Nitrogen fixation by common legumes (compiled from various sources)

Species	Fixation range (kg/ha/year)	Typical values suitable for calculation of nutrient budgets (kg/ha)
White clover (*Trifolium repens*)/grass mix (1)	75–450	150
White clover (*Trifolium repens*)/grass mix (2)	50–250	100
Red clover (*Trifolium pratense*) (1)	150–450	250
Lucerne (*Medicago sativa*) (1)	110–500	300
Vetch (*Vicia sativa*) (3)	100–250	150
Lupins (*Lupinus* spp.) (4)	100–375	100
Field beans (*Vicia faba*) (4)	200–350	250
Peas (*Pisium* spp.)	100–250	125
Green beans (*Phaseolus* spp.)	50	30

(1) cut for conservation, (2) grazed, (3) cut and mulched, (4) for grain (will vary between species)

Note that not all of these crops will be in place for a year. The effect of sward age on fixation is given in more detail in Table 4.4 below

Table 4.3 Management practices and their effect on N fixation

Practices which increase N fixation	Practices which reduce N fixation
Adding P and K	Adding manure
Cutting and removing material	Cutting and mulching
Mixing with grass*	Grazing
Short-term leys	Monocropping legumes
	Long-term leys

*As the proportion of clover in the sward declines, overall N fixation of the sward declines

Table 4.4 Estimates of nitrogen fixation by grass/clover swards of different ages (Kristensen et al., *1995,* Biological Agriculture and Horticulture *11: 203–219)*

Clover in sward (% ground cover)	Clover in sward (% dry matter)	N fixed (kg/ha/year)	
		Years 1–2	Years 3–5
10–29	3–16	80	47
30–49	17–29	157	84
>49	>29	248	128

Though leys give no direct economic return to a stockless producer, they do have other advantages. They reduce weeds in the subsequent crop, because they are strong competitors, while cutting reduces the weed seed-bank by preventing weeds from seeding. They also add organic matter to the soil and provide a break in disease cycles. Research indicates that it may be more profitable to bring a ley forward one year in the rotation rather than use a cereal break in vegetable rotations, because the cereal produces little income compared with vegetables and the ley produces a large benefit to subsequent vegetable crops, through reduced weed pressure/weeding costs, increased fertility and improved soil structure through addition of organic matter. These other factors also need to be factored in when deciding on the length of ley.

Green Manures and Cover Crops

The terms green manure and cover crop tend to be used interchangeably, while some leys could also be considered as green manures. They are grown principally for the benefit of the soil – retaining nutrients, adding organic matter and protecting the soil from erosion – though they may perform secondary functions, such as weed control, fodder production and disease breaks. Leguminous green manures are used to boost soil fertility through nitrogen fixation, often for a short period mid-rotation. Fixation follows the same principles as described above for legumes. However, as green manures are short-term crops, other species are often used. Vetch and red clover are commonly used in the UK, whilst other legumes, such as crimson clover (*Trifolium incarnatum*) and trefoil (*Medicago lupulina*), are used occasionally. Where they are used over the winter period, nitrogen fixation is likely to be small and leaving them in place for just a few weeks longer in spring can make a large difference to the amount of nitrogen fixed. They must be regarded as important crops and proper attention paid to seedbed preparation if they are to fulfil their true potential.

Non-leguminous green manures are used principally to retain nitrogen, which would otherwise be lost from the soil by leaching. They may also be used to give a short-term boost to soil organic matter or provide a break crop for other beneficial reasons. Cereal rye (*Secale cereale*) is the most effective species at nitrogen retention, as it establishes quickly and makes good growth over winter. It has a large, fibrous root system, which is good at scavenging soil mineral nitrogen and may retrieve nitrogen from lower down the soil profile, where shallow-rooted vegetables cannot access it. Mustard and rape and fodder radish are commonly used in arable systems, but they are less suitable for vegetable systems because of disease carry-over to brassica cash crops. Other, less common, green manures, such as phacelia (*Phacelia tanacetifolia*) and buckwheat (*Fagopyrum esculentum*), may be more suitable for vegetable systems. Phacelia is particularly effective at attracting beneficial insects, while buckwheat is able to effectively scavenge phosphorus from the soil, though neither is very frost-tolerant and both are more suitable for summer green manures. Whichever species is used, it is important to establish a good seedbed to ensure good germination and rapid establishment; this is particularly the case for winter green manures, which must be well established with a good ground cover before winter.

When it comes to incorporating green manures (and leys), consideration needs to be given to the effects on the following crop. Green manures, which are ploughed in early while still 'green', will tend to give a large, short, burst of nitrogen soon after incorporation. If the green manure is allowed to grow on towards maturity, then the material incorporated will release its nutrients more slowly, indeed there may even be an initial immobilization of nitrogen with non-leguminous crops. More mature crops will also have a longer-term effect on soil organic matter and structure. Sufficient time must also be allowed between incorporation and the next crop, because the decaying plant remains can inhibit germination of seeds. This has often been seen after rye and vetch, though other species can also be a problem. In vegetable systems where transplants are used this is not generally significant.

Crop Residues

Residues of cash crops can contain significant quantities of nutrients, particularly nitrogen, as well as useful quantities of organic matter. While residues of arable crops, particularly cereal straw, are low in nutrients but high in organic matter, the opposite tends to be true for vegetable residues. Table 4.5 gives some values for nitrogen content of some common crop residues.

Mineralization of Fertility-Building Crops and Crop Residues

The nitrogen contained in either fertility-building crops or crop residues must be released or 'mineralized' before it can be utilized by subsequent cash crops. Mineralization is a microbial process that converts complex molecules (proteins and the like) into simple inorganic ions (nitrate and ammonium), which plants can take up. The rate at which this happens depends on environmental conditions (especially temperature and moisture content of the soil), the microbial activity of the soil and the characteristics of the plant material. Crop residues with a wide carbon : nitrogen ratio will tend to release that nitrogen more slowly than those with a narrow ratio.

Table 4.5 Nitrogen content and carbon : nitrogen ratio of some common crop residues (with crop yield)

Crop	Crop Yield (t/ha)	Residue Total N (kg/ha)	Residue C : N ratio
Winter wheat	7.5	50	80
Spring barley	4.5	25	80
Winter oats	5.2	35	80
Early potatoes	25	85	23
Maincrop potatoes	40	100	—
Winter beans	3.5	80	—
Oilseed rape	2.2	115	—
Maincrop carrots	—	85–130	20
Onions	—	14	30
Brussels sprouts	—	90–180	18–22
Cauliflowers	—	50–100	—
Beetroot	—	200	13
Leeks	—	120	17

Information from various sources. Values are for conventional crops, organic crops will generally produce smaller yields and have a lower nitrogen concentration

Other characteristics of the material can be significant (for instance, large amounts of lignin slow the process down) but the C : N ratio is the most important factor in determining the rate of mineralization. Fig. 4.3 shows an example of the release of mineral nitrogen following the incorporation of vetch. Less is available after incorporation in March than in May because, by the later date, the vetch has had time to accumulate a greater biomass.

In order to make effective use of this nitrogen, its mineralization needs to be synchronized with the requirements of the following crop. If no crop is in place to exploit the released nitrogen it may be leached away, particularly in autumn, when temperatures are still high enough to allow decay of the residues. A cover crop can prevent leaching, retaining the nitrogen, which can then be released to the next crop in spring. With the exception of potassium, the return of other nutrients in residues is small. In the case of potassium, several hundred kg/ha can be returned to the soil and, as leaching of this element is negligible (apart from on very light sands), there is little need to worry about synchronization with the requirements of following crops.

The organic-matter content of crop residues makes little contribution to long-term soil organic matter, particularly the green residues from vegetables, as they are quickly decomposed by soil organisms. However, they do provide a useful boost to soil biological activity.

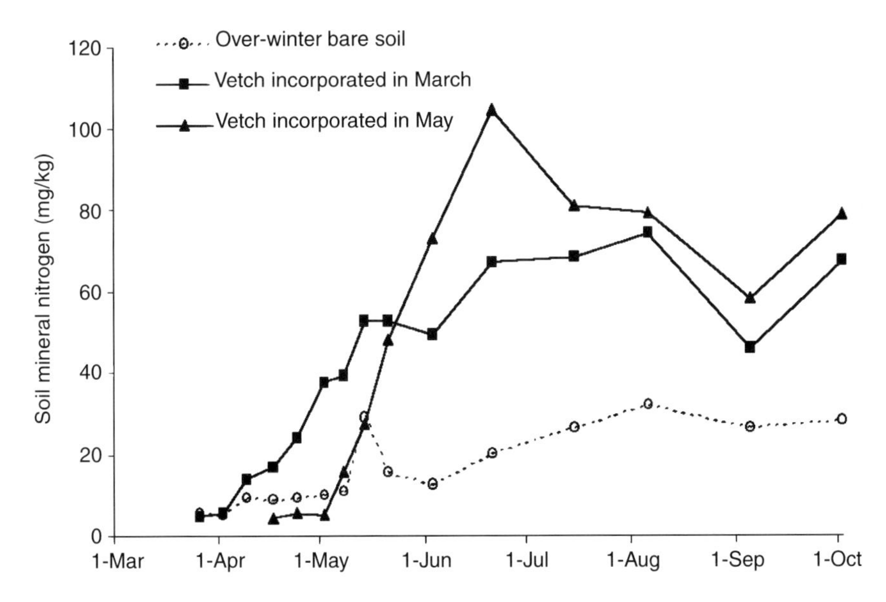

Figure 4.3 Mineralization patterns following either a winter vetch green manure crop or over-winter bare soil (Data collected by HDRA as part of DEFRA project OF0118T)

Animal Manures

Animal manures are the principal way of recycling nutrients on organic farms. They enable the recycling of nitrogen, phosphorus and potassium, together with other nutrients within the farm system. They can also be an important source of nutrients imported onto the farm in animal feed and bedding. In stockless horticultural holdings, imported manure often provides the main source of nutrients to balance the large off-takes associated with vegetable production. Manures also supply significant amounts of organic matter, which improve soil structure and boost soil microbial-activity.

Typical nutrient content of animal manures is shown in Table 4.6. The values shown are for total nutrient content and actual values will vary depending on animal diet, amount of bedding used, time of storage and so on. Manures from organic systems tend to be lower in the major nutrients, though this is not necessarily the case for the micronutrients. Not all of the nutrients in manure become available to the next crop. For farm yard manure (FYM) 15–24 per cent of the nitrogen is readily available, compared with 30–60 per cent for slurries. Phosphorus and potassium in manures become available within a year and can be considered as effective as conventional fertilizers.

Table 4.6 Typical nutrient content of manures and slurries

	Total N	**Total P(P_2O_5)**	**Total K(K_2O)**	**Total Mg(MgO)**	**Total S(SO_3)**	**Dry matter (%)**
Cattle FYM	6.0	3.5	8.0	0.7	1.8	25
	5.9	*3.1*	*6.6*	*1.6*	*2.3*	*25*
Cattle slurry (dairy)	3.0	1.2	3.5	0.7	0.8	6.0
Cattle slurry (beef)	2.3	1.2	2.7	0.8	0.7	6.0
Organic cattle slurry	*2.0*	*0.8*	*2.3*	*0.4*	*0.6*	*6.0*
Dirty water	0.3	Trace	0.3	—	—	1.0
Pig FYM	7.0	7.0	5.0	0.7	1.8	25
	6.5	*6.1*	*6.5*	*1.7*	—	*25*
Pig slurry	4.0	2.0	2.5	0.4	0.7	4.0
Sheep manure	6.0	2.0	3.0	—	—	25
Poultry manure	30	25	18	4.2	8.3	60

Numbers in italics refer to data from organic farms. Values are for fresh material. For solid manure they are in kg/t, those for liquid manures are in kg/cu.m
Values will be lower for manure stored outdoors for a long period, particularly for N and K
Values for slurries are typical and can be adjusted for actual dry-matter content
Source: ADAS Managing Manure on Organic Farms, *DEFRA RB209*

In order to minimize gaseous losses of nitrogen from animal manures, they should be incorporated as soon as possible. Table 4.7 shows the effect of delaying incorporation on nitrogen availability. Mid- to late winter and spring application of both manures and slurry will reduce nitrogen losses to the ground water by leaching, compared with autumn application. On grassland, application of solid manure should be avoided in summer, while slurry should be injected or a band spreader used in preference to broadcasting at any time of year.

Table 4.7 Amount of nitrogen lost from manures with different incorporation times

	Nitrogen lost	
Manure type	**10%**	**50%**
FYM	Immediate	6 hours
Slurry	1 hour	24 hours
Poultry	6 hours	48 hours

Source: ADAS

Aeration of slurry while in storage is recommended in organic systems, to reduce ammonia losses, reduce the viability of weed seeds and improve the handling qualities. However, aeration is expensive and incorrect operation can lead to a significant increase in ammonia and nitrous oxide emissions.

Composts

Composting is the biological process of converting organic material, including manures, into a relatively homogeneous stabilized product that can be used to build long-term soil fertility. Significant quantities of compost are now being produced from municipal and garden waste (1.66 million t in 2001), both by commercial companies and local authorities. These composts are often available cheaply and provide a valuable source of nutrients and organic matter, although strict limits on heavy metal content and GMO contamination imposed by the organic certification bodies mean that some of this might not be suitable for organic production. The organic certification bodies also recommend that animal manures produced on the farm are composted before spreading, while an increasing number of farmers are involved with on-farm composting of mixed organic waste brought onto the farm.

The composting process can be broken down into a number of relatively well-defined phases, though there is inevitably some overlap in the processes occurring. There is an initial rapid heating of the pile up to around 60°C, as micro-organisms (mainly bacteria) breakdown readily decomposable compounds, such as starches, sugars, fats and proteins. This phase lasts two or three weeks and it also serves to kill weed seeds and pathogens. As the readily decomposable compounds are used up, the decomposition rate slows and pile temperature begins to fall. Different organisms (including fungi) take over the degradation of more resistant compounds such as cellulose. Finally, fungi come to dominate the pile and begin to break down the most resistant compounds, such as lignin, though this is a very slow process and, at this point, the composting process is effectively over, with the pile temperature similar to ambient air-temperature.

For successful composting to occur, the pile is usually turned several times, in order to maintain aerobic conditions. In the initial phase, turning should be at least once a week, but later on can be at longer intervals. Failure to turn the pile leads to anaerobic conditions, in which the whole decay process slows down. Toxic compounds can also be produced. Turning out manure from a shed into a field and leaving it there for six months is *not* composting. In commercial situations, static piles with air forced in through perforated pipes and enclosed composting vessels are also used, though capital outlay is high and these methods are not

Turning compost with tractor-mounted windrow

really suitable for on-farm composting. As well as turning, for successful composting the initial mixture must be suitable. Table 4.8 gives an indication of the main factors to consider, while Table 4.9 indicates how this starting mix can be achieved. Failure to control these factors can lead to slow or incomplete composting or problems with odours, dust and the like.

Table 4.8 Main factors to control in preparing material for composting

Factor	Acceptable range	
Carbon : Nitrogen ratio	20 : 1–40 : 1	↑
Moisture content	50–65%	Increasing importance
Particle size	1–1.5cm	
pH	6.0–9.0	↓

Table 4.9 Typical C : N ratio and moisture content of common compost ingredients

Material	C : N ratio	Moisture
Vegetable crop residues	13 : 1–30 : 1	75
Cattle FYM	10 : 1–30 : 1	65–90
Poultry manure	3 : 1–10 : 1	60–75
Grass clippings	17 : 1	82
Straw	50 : 1–150 : 1	5–25
Sawdust	200 : 1–750 : 1	20–25

During composting, much of the organic carbon is respired by the organisms doing the composting. What remains is a mixture of resistant compounds left from the initial organic matter and highly complex molecules formed by the continuous re-combination of organic matter during the decomposition process. These latter compounds are similar in nature to soil humus and contain stabilized forms or organic nutrients, including most of the nitrogen, sulphur and a significant proportion of the phosphorus.

In summary, the advantages of composting can be stated as:

- mixed organic material, such as manure, straw, crop residues, waste wood, green garden waste, food-processing waste, are converted into a homogeneous product suitable for building long-term soil fertility
- effectively managed aerobic composting destroys weed seeds and pathogens
- product volume is reduced, increasing ease of handling.

Offset against this, the disadvantages of composting can be seen as:

- more than half of the carbon and nitrogen can be lost to the atmosphere during the processes and significant potassium (more than half) and some nitrogen and phosphorus can also be lost by leaching
- the nitrogen in compost is stabilized and so not readily available to the crop (as a rule of thumb, less than 10 per cent of the total nitrogen will become available in the first year after application as compared with 10 per cent for autumn application and 25 per cent for spring application of fresh FYM)
- composting is time consuming and there may be significant initial capital outlay, depending on the scale of the operation.

Application of compost is done on the basis of total nitrogen content, as with manures. Up to 250kg per ha total nitrogen may be applied per year under the Code of Good Agricultural Practice for the Protection of Water, or 500kg per ha total nitrogen every two years in catchments insensitive to nitrate leaching. At this rate, one-off applications of compost have little effect on crop yields, as scarcely any of the nitrogen is available. However, a significant proportion of the phosphorus and potassium is available and can contribute to crop nutrition. This is especially useful in a horticultural rotation, because of the high phosphorus and potassium demand of many crops. Composts produced commercially from organic wastes also contain significant micronutrients, such as copper and zinc, which are required in large amounts by some horticultural crops. The relatively high phosphorus and potassium content of some composts means that they should not be used principally as a nitrogen source, as this can lead to a build up of excessive phosphorus and potassium.

As well as supplying nutrients, composts add useful organic matter to the soil, though perhaps not as much as might be expected. Typically, compost contains 20–30 per cent organic matter, but some commercial composts may contain even less than this. However, unlike manure, the organic matter is stabilized, only breaking down slowly in the soil. As a result, regular applications can quickly build up soil organic matter, improving soil structure. This is especially important in vegetable systems, where intensive cultivations, often on vulnerable soil types such as silt, combined with harvesting operations in unsuitable ground conditions enforced by marketing constraints, make soil structure a regular problem.

Supplementary Nutrients

A farm system that is continually exporting nutrients off-farm will require these to be replaced, if a long-term decline in soil fertility is to be avoided. This is particularly true in the case of horticultural systems, as many of the crops are high in nutrients with little retained on-farm for animal feeds. Particularly rapid declines of available soil phosphorus have been reported after conversion to organic vegetable cropping, when insufficient supplementary nutrients have been supplied. Sulphur deficiency is also likely to become more widespread, particularly on light soils, as low-sulphur fossil fuels become the norm. Atmospheric deposition of sulphur fell from more than 70kg per ha per year over most of the UK in the mid-1970s, to less than 10kg per ha per year by the late-1990s. Response to sulphur fertilizers has been demonstrated widely in conventional systems, not only in brassicas, which require a large amount of sulphur, but also in cereals. Deficiency in crops looks similar to nitrogen deficiency and, as well as reducing yields, it also reduces nitrogen fixation by clover. In many cases, brought in animal feed or bedding will provide enough supplementary nutrients. In other cases, manure or compost may be brought in to balance nutrient removals in crops. In some cases, however, other forms of supplementary nutrients will be required.

Organic standards permit an eclectic mix of organic and non-organic materials to be used as fertilizer in organic systems, from wood ash and seaweed extracts to ground-up rocks and waste products from industry and even, in some cases, standard micronutrient fertilizers. The use of some of these products is restricted, requiring evidence of a need before certifying bodies permit their use. This can usually be provided through soil or crop analysis.

In order to assess if supplementary nutrients are likely to be required, nutrient budgeting should be carried out for macronutrients, along with regular soil testing (*see* below). Nutrient budgeting is of little use for micronutrients and deficiency is often more difficult to detect. Deficiencies prior to conversion should give clues, though specific soil or crop testing, in concert with advice from an agronomist, will be required if micronutrient deficiency is suspected. Such tests are required by certifying bodies before some micronutrient fertilizers can be used.

Physical Management

Attention to the physical structure of the soil is as important as attention paid to the nutrient status of the soil or its biological activity. Physical structure is important for a whole range of factors, including water and air movement, seed germination and root penetration, nutrient availability and soil erosion. Poor physical structure can lead to a host of problems, such as increased pest and disease pressure, reduced water and nutrient availability, reduced soil trafficability and soil erosion. Structure is particularly important in vegetable-based systems, as many crops are vulnerable to poor structure because of their small, weak, root systems and their vulnerability to pest and disease pressure. The intensive cultivations associated with seedbed production for small-seeded crops, such as carrots, are also very disruptive to soil structure. Time windows for crop-management operations also tend to be smaller than for arable crops, meaning they may need to be carried out while soil conditions are not ideal. While a conventional farmer can compensate for poor structure to some extent by increasing use of fertilizer and biocides, these options are limited in an organic system and do not fit well with the organic philosophy of managing the whole system.

Vegetable production has traditionally been concentrated on sandy and silty soils, which are also the most vulnerable. Structure is delicate and they are easily overworked, leading to slumping and surface capping, particularly in silts. Signs that soils may be suffering structural damage include fields slow to drain and quick to waterlog during rain, poor crop growth (often due to nutrient deficiency), increased pest and disease pressure and reduced workability of the soil. Often, these symptoms are patchy and concentrated on headlands or other areas with high trafficking rates. If soil structural problems are suspected then an assessment can be made using the method described below.

Soil structural problems can be difficult and expensive to correct, particularly on heavy soils, and so they are best avoided in the first place. Cultivating or trafficking when the soil is too wet, particularly on heavier soils, is the easiest way to damage soil structure. Under such conditions soils smear and clay particles can be disrupted, blocking air spaces and forming hard impermeable clods as the soil dries. Over-cultivating a soil, particularly with power cultivators when the soil is dry, can also cause problems, as the dust produced can be washed deep into the soil profile,

Soil structural problems can be caused by excessive field traffic in wet conditions

clogging pores and impeding drainage. These fine particles may also be lost in drain water, taking nutrients with them and damaging water courses. As a result, timeliness of operations is of paramount importance. Options for reduced or minimal cultivation are limited in organic and vegetable systems. However, it is generally best to reduce the amount of cultivation as far as possible, by reducing passes and running power cultivators at a lower speed. This is especially important when the soil is being worked in less than ideal conditions and it also reduces establishment costs. Once the soil has been cultivated it is very vulnerable to compaction, so trafficking should be kept to an absolute minimum. This is especially the case after subsoiling. Using low-pressure tyres or cage wheels can reduce compaction problems, but their effectiveness declines as the soil becomes wetter and they can even increase the area of soil damage, once the soil reaches a point where any trafficking causes damage. Much vegetable production is done using a bed system, which concentrates compaction by machinery. However, during harvesting operations it is common to see tractors driving indiscriminately over the field. This can cause widespread soil damage (*see* Table 4.10), particularly if soil conditions are poor. This is especially important if harvesting a cereal break crop that has been under-sown with a ley, as there will be limited options to remedy

Table 4.10 The effect of the number of tractor passes on degree on compaction

Number of passes	Proportion of compaction caused
1	50%
2	10% more
3	6% more
4	3% more

Data from A Guide to Better Soil Structure, *Cranfield University*

damage. Leys with poor soil structure will have reduced growth and nitrogen fixation, reducing yields in subsequent crops. Harvesting operations should use existing wheelings as far as possible, which can be worked with tines or subsoiler afterwards, to correct compaction, even in a ley.

If animals are to be grazed on a ley, poaching should be avoided by controlling stocking rates and limiting damage in vulnerable areas by regularly moving feeding- and watering-troughs. Where pigs are grazed on a ley, it is important to maintain a cover of grass, otherwise serious damage to soil structure can occur. Though pigs can be used to clear ground of perennial weeds, this should only be done when absolutely necessary.

Improving Soil Structure

As well as avoiding activities that damage soil structure, much can be done to improve soil structure, to make it more robust, giving wider time-windows for 'safe' cultivations. The best way to do this is by adding organic matter. Organic matter provides food to soil organisms, which help bind the soil minerals into aggregates, while the humus, produced by organic-matter breakdown, possesses an electrochemical charge that binds to clay particles, again helping bind soil-aggregates together.

Many soils used for vegetable production have low soil organic matter, particularly the lighter soils, because of intensive cultivations, which expose organic matter to micro-organisms that then break it down. Many systems are stockless, with little or no animal manure inputs, particularly in the east of England, and while there can be a significant amount of crop residues produced, particularly from brassicas, these are high in nutrients and decay rapidly.

Organic vegetable systems, particularly where they are converted from intensive conventional systems, should make increasing soil organic matter a priority. Ideally, this will involve introducing animals into the system to provide manure and allowing for the economic use of longer leys. However, this may not be practicable if the infrastructure and knowledge required are absent. In this case it will be necessary to import organic matter onto the farm. This can be animal manure from another holding. Ideally this will be an organic holding, best under a reciprocal arrangement, where animal fodder, such as vegetable outgrades, is exchanged for manure. If the manure is obtained from a conventional holding it will require some treatment before use. There is also the possibility of importing either prepared green-waste compost or organic wastes, which can be composted on farm. In areas with extensive vegetable production, there may be significant quantities of vegetable-processing waste from factories and pack houses available for producing compost on farm.

The other way to increase soil organic matter is through growing leys. However, in stockless or low-stock situations typical of vegetable production, time in ley must be kept to a minimum, as they provide no direct income. This may be overcome by renting out grazing to other farmers, but the possibilities for this may be limited, particularly where conventional stock are concerned. The improvement in soil structure associated with incorporation of a ley is also relatively short-lived, generally lasting only two years. An alternative strategy to the traditional use of one, two or three year leys, followed by two, three or four years cropping, is to use lots of short-term green manures and cover crops to minimize periods between leys and maximize organic matter inputs (*see* below).

Nutrient Budgeting

Achieving a balance between nutrients exported from the farm in produce sold and nutrients imported, in animal feed, bedding and manure/fertilizers, is important, both for short-term productivity and long-term sustainability. Consideration must also be given to individual-field nutrient-budgets, as wide variations can occur within a whole farm that is in overall balance. There are a number of methods of compiling nutrient budgets (*see* Table 4.11 for an example). The more complex methods are more accurate, but require more data, which may not be available. The boundaries used for the budget, both physical and temporal, must be defined and will, to some extent, be determined by the purpose of the budget. For example, if rock phosphate is applied once per rotation, then annual budgeting of phosphorus may be meaningless. In general, whole rotation budgets are needed for organic systems because of the distinct fertility-building and exploitative phases.

It is normal to produce nutrient budgets for nitrogen, phosphorus and potassium, as these elements are likely to see the largest fluxes and are most likely to limit crop production. Determining nutrient inputs and outputs is

Table 4.11 Simple field-scale nutrient budget for a field vegetable rotation using undersown barley as a break crop and as a means of establishing the fertility-building phase of the next rotation

Year 1	Year 2	Year 3	Year 4	
Grass/red-clover ley (cut and mulched)	Potatoes (followed by rye cover-crop)	Onions	Spring barley (undersown)	
Nitrogen				
+20kg/ha (deposition) +250kg/ha (fixed) + 205kg/ha (in compost)	+20kg/ha (deposition)	+20kg/ha (deposition)	+20kg/ha (deposition)	Total gains: 535kg/ha
–25kg/ha (leaching)	–60kg/ha (leaching) –75kg/ha (in 29t/ha yield)	–60kg/ha (leaching) –33kg/ha (in 19t/ha yield)	–60kg/ha (leaching) –26kg/ha (in 2.1t/ha yield)	Total losses: 339kg/ha N balance: 196kg/ha
Phosphorus				
+27kg/ha (in compost)				Total gains: 27kg/ha
	–22kg/ha (in 29t/ha yield)	–8kg/ha (in 19t/ha yield)	–7kg/ha (in 2.1t/ha yield)	Total losses: 37kg/ha
				P balance: –10kg/ha
Potassium				
+116kg/ha (in compost)				Total gains: 116kg/ha
	–162kg/ha (in 29t/ha yield)	–41kg/ha (in 19t/ha yield)	–11kg/ha (in 2.1t/ha yield)	Total losses: 214kg/ha
				K balance: –98kg/ha

This rotation has an apparent surplus of nitrogen but a deficit of P and K. However, much of the nitrogen surplus is derived from the compost application and this will be largely in an unavailable form for some time. Different cropping will affect the budget – for example, growing carrots rather than onions would make the P and K balances more negative (a 42t/ha crop of carrots could remove 48kg/ha N, 18kg/ha P and 152kg/ha P).
Green-waste compost is used as the only externally sourced soil amendment
Values are based on data collected by HDRA on a sandy loam in central England as part of the DEFRA project OF0191

relatively straightforward in some cases, for example with supplementary nutrients, but very difficult in others, say the amount of nitrogen fixed by a ley. For maximum accuracy, the nutrient content of crops, manures and so on should be measured, though in practice there are standard values available. Most organic systems show a surplus of nitrogen and phosphorus, with potassium sometimes in surplus, sometimes deficit. However, horticultural systems are usually in surplus for all macronutrients, often significantly so, because of the widespread use of purchased manures and fertilizers. The long-term sustainability of such systems is questionable, as nutrient accumulation will eventually lead to losses to the environment, with consequential pollution effects. It also illustrates the problems of trying to manage nutrients in organic systems, where the sources available, such as manures and composts, do not contain nutrients in the required ratios. Nutrient budgets are not a substitute for soil analysis, but they can provide a useful tool for judging what is happening in organic farming systems. One significant drawback of them is that they do not show when the nutrients will be made available. In the case of nitrogen, for example, it is important that this happens at the times when a crop demands it.

Soil Analysis

This is an important practical tool in the management of soil fertility for any grower, as a means of ensuring that the soil is in good condition and particularly to pick up trends over rotations and longer time periods.

Testing for Chemical Fertility

Testing for plant nutrients should be carried out twice per rotation, ideally at the same points in each cycle. This will indicate if soil nutrients are being accumulated or depleted, although the spatial variability of soil means that slight changes will not be picked up from one sample to the next.

In order for soil testing to be effective, the sampling must be representative of the whole field. Poor sampling can lead to expensive errors in use of manures and fertilizers. Sampling should be done in the autumn or spring, when the soil is not too wet, but not within six months of lime, manure or fertilizer application. Samples should be collected with an auger, as this will ensure that the sample contains equal amounts of soil throughout the sample depth. In arable and horticultural fields, sampling should be to around 15–20cm (6–8in) and in long-term grassland 7.5–10cm (3–4in). Take around twenty-five samples from a uniform area. If the soil in a field varies significantly, as for instance along a slope, then the different areas must be sampled separately. To get an even sample, walk over the area to be sampled in a large 'W' pattern, taking samples at regular intervals. Areas with unusual features, such as gateways, feeding troughs and trees, should be avoided, as should headlands. Once the sample is collected it should be thoroughly mixed and a sub-sample placed into a clean, clearly labelled bag before being sent for analysis. Around 3kg (6.5lb) is required.

There are a number of testing options available to organic farmers. In England and Wales, the standard ADAS soil testing provides soil indices for phosphorus, potassium and magnesium, as well as pH and total organic matter, as standard. In Scotland, slightly different methods are used by the SAC testing service, reflecting the different types of soils in Scotland. Using these methods has the advantage that the indices will provide continuity for comparisons, as farms convert from conventional through to an organic system. The indices have also been tested against performance of a wide range of crops over a long period. They are, however, not foolproof in predicting crop performance and may be less appropriate for organic systems.

The second option is to use the soil-testing service provided by the Organic Advisory Service of Elm Farm Research Centre. This service uses different methods, which are said to be more applicable to organic systems. However, there has been little testing to show how the results produced relate to crop performance. This is especially the case with horticultural crops. The values produced will still be useful in determining trends in soil nutrients if used consistently.

Finally, there are a number of laboratories in the UK offering soil analysis based on the ideas of William A. Albrecht, Professor of Soils at the University of Missouri in the early twentieth century. This analysis places great emphasis on the chemical balance of nutrients in the soil, particularly the importance of calcium and also of organic matter. The testing provided is very comprehensive and comes with detailed interpretation, but is expensive.

As different systems of analysis use different methods of measuring the same thing, they produce different results for the same soil. It is not, therefore, possible to make a direct comparison between results obtained from these three methods. However, consistent and systematic use of the same method will enable the overall trends in soil nutrients to be evaluated as discussed above.

Testing for Biological Activity

Though biological activity in the soil is vital for the breakdown of organic matter, the release of nutrients and the control of soil-borne pests and diseases, it is very difficult to measure in a routine way. However, an active earthworm population and rapid breakdown of crop residues and manures is a good indicator of good soil biological activity. Some laboratories are beginning to offer tests, but these probably still need development to be of immediate practical benefit to growers.

Testing for Soil Physical Conditions

Assessing soil physical characteristics is important for the many reasons described above and several methods for doing this are briefly described below.

Soil Texture: before assessing the soil structure it is important to know its texture. If you do not know this it can be assessed by a testing laboratory, though using the method below will give an accurate enough assessment for most agronomic purposes.

Take a small sample of soil and moisten it if required so that the particles hold together. Work it between fingers till it is consistent. The guide in Table 4.12 allows a rough interpretation of soil types, based on the ADAS texture classes, which are a simplified version of the twelve texture classes (*see* Fig. 4.1).

Soil Physical Conditions: there are a wide range of laboratory-based methods that assess soil physical conditions. However, for agronomic

Table 4.12 Soil texture characteristics when working a moist soil bolus

Soil type	Characteristics
Sandy	Will not form a ball. Not sticky. Gritty feel
Loamy sand	Can be moulded when wet, but falls apart readily. Does not stick to fingers. Gritty feel.
Sandy Loam	Can be moulded into a weak ball that can be deformed without falling apart. Feels gritty.
Loam	Easily moulds into a ball, particles bind together well. No other predominate features.
Silty loam	Moulds easily, with a silky feel. Sticks to fingers.
Clay loam	Binds together strongly into a ball which resists deformation. Forms a polish when smeared between fingers and is sticky.
Clay	Binds into a strong ball that is difficult to deform. Forms a high polish when smeared between fingers.
Peat	Very dark in colour. Forms an easily deformed ball.

There are sub-classes within these types. In the case of the sandy soils it depends on the size of the sand grains. For example, a sandy loam in which the sand is fine is a fine sandy loam, in the silts and clays it depends on the presence of other size classes. For example, a clay soil which has a gritty feel is a sandy clay, a clay loam with a silky feel a silty clay loam. With peat it depends on the presence of a mineral fraction. For example, a peat which has a gritty feel is a sandy peat.

purposes, soil physical condition can be assessed relatively easily in the field. Regular assessment will show if problems are developing, allowing remedial action to be taken at an early stage. All that is needed for assessment of structure is a spade.

Push the spade in on three sides of a soil spit; on the third side remove the spit and lay it on the ground, with the fourth, un-dug, side facing up. Use a knife to gently prise apart the soil and examine the structure. If it is not clear how the soil is structured it can be useful to drop the spit so that the soil breaks up along its natural lines of weakness. Structure below the plough layer can be examined by enlarging the pit and digging a spit from deeper in the soil profile. Careful examination of the excavated spit can tell you a great deal.

Compacted zones – Probing the soil with a knife, pencil or such like will help show up areas of compaction. They will tend to have large clods, which are resistant to pressure between the fingers.

Clods – Large clods with smooth surfaces indicate poor structure. Small clods with irregular surfaces (usually known as crumbs) indicate good structure. No consolidation into clods at all indicates poor structure (sandy and silty soils).

Cracks – Cracks contain air for roots and soil organisms, allow drainage and provide channels for root growth. Many cracks of various sizes indicate good structure, as opposed to few cracks that indicate a poor structure. Predominately horizontal cracks are indicative of compaction.

Roots – In soil with good structure, roots will be numerous and proliferate through the whole profile. If they are restricted to major cracks, balled up or growing sideways there are problems.

Anaerobic conditions – Poorly drained and/or compacted soil can develop anaerobic regions. They may contain black un-decomposed crop residues and may smell foul. Long-term anaerobic conditions are more likely in the lower part of the soil below the plough layer. Under these conditions a grey colour can develop and even a sulphurous smell.

Biological activity – Poor structure inhibits biological activity. Numerous earthworms and their channels, other soil life and rapid decomposition of crop residues are indicative of few structural problems.

The assessment should be done when the soil is moist, preferably when a crop is growing, to show roots, and well after the last cultivation. At least ten such spits should be examined in a field, more if there is a clear difference across the field due to slope or some other physical feature. Areas prone to compaction, such as gateways and headlands can show up problems early. As soil type will affect the appearance of the soil, it is useful to have an experienced advisor to help with initial assessments. However, as a rough rule of thumb, the more clay a soil contains, the larger will be the clods and the fewer the cracks, so a well-structured clay soil will look different to a well-structured sandy loam. Often, a soil may have good structure in the upper layers that undergo regular cultivation, while lower layers may be compacted, so it is important not to concentrate on the cultivated layer only. Areas of permanent grassland generally have very good structure, but temporary leys can suffer poor structure and so also need to be assessed.

Chapter 5

Weed Management

Weed management can be defined as any technique that favours the crop over the weeds. Whilst the costs of weeds in terms of both yield and economic loss are generally well understood, the beneficial properties of weeds also need to be taken into consideration. These might include such factors as prevention of leaching, erosion control and refuges for insect predators of crop pests. In order to balance these factors, and in keeping with organic principles, organic growers should not be aiming to completely eradicate weeds from their farming systems but to manage them. Weed management, that is maintaining low, beneficial and tolerable levels of weed infestation, should be the goal in any weed-control programme. However, this may make organic weed-management programmes more complex and long term than those practised in chemical-based systems.

The following chapter aims to outline the methods available at present to help achieve this in organic horticultural systems. This chapter moves logically through the weed-management decisions that face farmers and growers in a practical situation. Preventative or cultural control techniques are considered first, followed by a review of farm hygiene and, finally, the tactical decisions that might need to be taken in the post-establishment phase. In the former case, the importance of taking weed management into account in the planning phase of the system is stressed and fully discussed.

CULTURAL CONTROLS

Crop Rotation

Crop planning is a cornerstone of organic farming practice and is integral to the whole systems approach, so it has important implications for weed management. It can be designed to positively influence weed control and to make a useful contribution to the whole-farm management strategy. Elements of the design strategy for crop rotations are considered separately below.

Crop Choice and Sequence

The length of the rotation, the choice and sequence of crops will depend upon individual farming circumstances that will include factors like soil type, rainfall, topography and enterprises. However, it is desirable to produce an unstable environment in which no single weed species is allowed to adapt, become dominant and, therefore, impossible to manage. No one rotation can be recommended but, ideally, in terms of weed control rotations, it should include:

- alternation of autumn- and spring-germinating crops
- alternation of annual and perennial crops (including grass)
- alternation of closed, dense crops, such as oats, which shade out weeds, and open crops, such as maize, which encourage weeds
- a variety of cultivations and cutting or topping operations that directly affect the weeds.

In addition, consider putting sensitive annual crops after perennial leys. This is because it has been shown that in the third cropping year after a grass/clover ley there can be twice as much weed emergence as compared to the first.

Fertility-Building Leys

The length of the fertility-building period within the rotation will have an impact on the weed population. If managed well, the ley period can act as a weed-suppressing phase. The choice of fertility-building crop and ensuring good establishment are both important. Rotations with grass leys have been shown to be beneficial in reducing weed-seed numbers compared with rotations that do not include a grass phase. Grassland systems that have temporary leys rather than permanent pasture will provide the opportunity to control perennial weeds during the cultivations between ploughing and reseeding.

Establishment of leys can be easier in the autumn period than in the spring, because sowing in spring coincides with the main spring-flush of weeds. The seedbed needs to be well prepared and good contact made between seed mix and, ideally, moist soil to achieve good establishment.

Cover or Break Crops

Introducing cover or break crops in the rotation can also be important for weed management. Break crops are so-called because they 'break' the cycle of normal cropping and provide diversity in rotations. This change of cropping may have several benefits: pest and disease control, nutrient use efficiency, higher economic returns and also improved weed-management. For example, potatoes are a broadleaf smothering crop with different growing methods and different associated weed-flora compared with, for example, a cereal crop. Potatoes will also require the use of different weed-control methods, providing an opportunity to cultivate the soil with alternative implements as compared to the cereal.

Cover crops have the primary purpose of nutrient management, but they also tend to be quick germinating and dense, so that they suppress weed emergence. Some stockless systems rely on short bursts of green manures rather than leys for fertility and will lose the benefit of a longer weed-suppressing phase. This means that it is essential to get good establishment of a dense crop to prevent weed problems. Some cover crops may also exhibit allelopathic properties, that is they exude chemicals that can have an inhibitory effect on surrounding plants. For example, when grazing rye has been incorporated and is breaking down, it has been shown to reduce the emergence of weed seedlings in a subsequent crop.

Intercropping and Under-Sowing

Two or more different crop species can be grown together. This is known as mixed or intercropping. The advantage of this practice, in terms of weed control, is that there is greater ground coverage by the crops, leaving less area available for weed emergence. The mixed cropping can involve purely cash crops, or a mixture of cash crops and fertility-building crops. These can be arranged as under-sown crops, strip crops or row intercrops. In terms of mixed cash-cropping there have been investigations into organic winter wheat and beans, which reduced weed growth and gave better yields than sole cropping. Leeks and celery intercropping has also been shown to increase weed-suppressing ability and reduce reproductive capacity of late-emerging groundsel.

Under-sowing reduces weed emergence, as a quick-growing dense layer of vegetation covers the ground. If cash crops are under-sown with fertility-building crops, then there can be advantages other than weed suppression and yet this method has only been widely adopted for use in cereals, although there are examples of more novel vegetable-production systems using under-sown legumes. If the under-sown prostrate species is leguminous, this can potentially add to or maintain soil-nutrient status. If cash cropping is combined with fertility building in this way, then an economic return can be achieved and there may be no need for a dedicated fertility period, or at least the length of that phase can be reduced. However, for vegetable-production systems, a lot more field experimentation is needed in variety choice and sowing rates, to ensure competition between the two crops is kept to a minimum.

Variety Choice and Seed Quality/Grade

Some varieties can be said to tolerate weeds, whilst others actively suppress them. Organic systems need varieties that actively suppress weeds, if possible. Quick germination, good early vigour, large leaf-area, prostrate growth or height, are all desirable traits that will shade weeds. Spatial distribution of the canopy foliage and rooting system will also be important. Some varieties seem to exude allelochemicals into the soil, which work to suppress weed development. Growers should try and choose vegetable varieties that display at least some of these weed-suppressing characteristics.

Varieties with a larger seed size have also been shown to exhibit greater initial vigour of emergence and growth, which may subsequently provide extra competitive ability. If there is a choice available, then the most vigorous species should be selected, as these will be more likely to outcompete weeds and suppress their development.

It is important that the crop seed is free from contamination by weed seeds. Organic growers are obliged to use organically produced seed and this should be clean. In some instances, tolerance levels for weed-seed presence may exist. If growers are saving their own seed it is important that they take it from weed-free crops and, ideally, have it professionally cleaned.

Seed Rate and Crop Spacing

In drilled or transplanted crops the proximity of the plants to one another will determine the competitiveness of the plant stand as a whole. Seed rates tend to be higher for organic than conventional crops. The theory is that the greater the amount of space taken up by the crop in the rows, the less space there is available for the weeds to invade. There is also an allowance for potentially lower germination-rates and loss of the crop by mechanical weeders. There has been much work in cereals on row spacing, pattern, direction of sowing and seed rates. Results are varied and interactions between the factors are often significant. For example, some research has shown that in narrow row widths an east/west sowing was favourable, whereas in wide row-widths a north/south sowing showed better response. Similar considerations should apply, in principle, in vegetable crops, but the work done in this respect is limited. In terms of row spacing it maybe that market size-specifications are the main drivers in determining the crop spacing.

Crop Establishment

The ability of the crop to get off to a good start ahead of the weed flora is critical. This can be aided by the use of primed seed or by transplanting an already established plant into a freshly prepared weed-free seedbed.

Transplanting is a popular technique in organic horticultural systems. Bare-rooted transplants can be raised on holdings or modular plants raised or bought in then planted out in the field. This has the benefit of accurate spacing, namely not having to rely on germination, which can sometimes lead to uneven establishment with subsequent yield and quality penalties. It also accentuates the difference in size between crop and weed, which can be vital for mechanical weeding at later stages.

Cutting Regimes

In horticultural and stockless arable systems, ley management will include topping at intervals during the summer to a height of around 10–15cm (4–6in). Ideally, in fertility-building leys, the sward should not be allowed to get higher than 40cm (16in or knee height). If the vegetation gets higher than this, then topping will create a mat of vegetation that will act like a mulch. This can create dead spots in the ley, where clover may be excluded by the more vigorous grasses or which weeds may colonize. Topping the ley regularly will also ensure that tall weeds that may have germinated will not be able to set seed. In grazed pasture, weeds that are not eaten by livestock will need to be topped to prevent seed shed. Topping can also be used as a remedial measure in vegetable crops to prevent weeds from seeding.

In grassland systems, cutting for hay or silage will have an impact on the weed flora. Silage tends to be cut early in the season when the sward is young and fresh, whilst hay is cut at a later stage. There can be both advantages and disadvantages associated with the timing of cutting, depending on the weed flora and the ultimate requirements of the system. Cutting late may allow weeds in the pasture to grow to maturity and set seed. The ripe seeds may contaminate the hay and remain viable when passed through livestock. Dock seeds should not survive low-pH silage, however, they will survive in a later cut of hay. This mature seed may also shed on the ley surface and find opportunities to germinate *in situ* or be transported by livestock to other locations. In contrast, cutting early for silage in fields with, for example, an infestation of creeping thistle may encourage the spread and growth of this weed. Hence, there has to be a balance between the requirements of the farming system and weed-control implications.

Use of Manures

Placing manures and slurry more accurately on crops can work to benefit the crop rather than the weeds. In arable/horticultural systems, if manure is placed 10cm below the soil surface, the deeper-rooting or deeper-sown crop seeds are encouraged to grow down into the nutrient-rich layer in preference to the weeds. It has also been shown that if slurry is being broadcast spread it is better for it to be ploughed in than left on the surface, in order to favour the crop. This is because the crop tends to germinate from deeper in the soil profile, accessing the nutrients before the more surface-dwelling weeds, which tend to germinate from 0–3cm depth. After silage cuts, the application of slurry to stubble may provide optimum conditions for weed seed germination. This nutrient-rich bed of cattle slurry will produce a high-potassium environment that will favour weeds such as docks over grasses.

Livestock

In mixed systems, where grass/clover leys are used for fertility building, livestock can make good use of the nutrients and produce a resource that can be used around the farm to fertilize cash crops. Management of the ley becomes important in livestock systems. This includes getting the right grazing balance over the year, namely lighter stocking in winter months and in wet periods to prevent poaching of the area and, ideally, alternating (from year to year) with sheep/cattle or mixed grazing. Grazing rotations will help, as sheep may graze out certain weeds, say ragwort, whilst free-range outdoor pigs forage for roots and could reduce the perennial weed burden. Sheep may also be useful to lightly graze oats to encourage tillering and, hence, weed suppression.

Fallowing

It is not desirable to have to plan a fallow period into a rotation, but it may be necessary if weeds cannot be controlled during cropping or fertility building. It may not be necessary to stop cropping for a whole year but, instead, to employ a bastard fallow – that is, no crop for part of the year. This is often in the summer, when cultivations can take place and the drier periods allow for root desiccation. This technique is more useful in plough-dominated systems rather than grassland management. Perennial weeds are usually more of a problem in grassland and have a capacity to regenerate and spread vegetatively as well as by seed. In theory, cultivation could keep them in check, but much more research is needed into machinery, time, frequency and height of cutting, removal methods and also prevention techniques.

FARM HYGIENE

Hygiene is an important part of any weed management approach. This prevents the spread of infestations around the farm and is often overlooked when planning weed-management programmes. Some of the principal issues are discussed below.

Inputs

It is essential that any compost, manure or slurry that is to be spread on fields is free from viable weed seeds. This means that, if compost is made on the farm, the heap will need to be turned to ensure the temperature rises high enough to kill any weed seeds that may be present. The temperature needs to remain at 55–65°C for seven or more days (in windrows) to ensure the heating process has been effective. Ideally, the compost or manure should be covered to prevent the introduction of additional wind-borne seeds, or at least kept weeded to ensure no seeds germinate and set seed in it. Slurry should be adequately aerated to ensure weed seeds have been killed.

Farm Machinery

If there is a serious weed infestation in a particular field or machinery is moving through fields with flowering weeds, then washing-down machinery should be considered. This may be particularly important at harvest, where some weeds may be setting seed in fields. This also applies to contractors coming onto the farm who are able to transport weed seeds between holdings. For crops like cereals, where weed seeds may be scattered in the field or caught on the machinery and dislodged later some distance from the original source, it may be necessary to think about screens on combines to catch weed seeds at harvest. Older models may already have these features.

Management of Non-Cropped Areas

Areas of the farm that are not cropped still need to be managed to prevent the spread or invasion of weeds. It has been shown that some perennial weeds from field margins are less invasive if a strip of grass or wildflowers is sown around the field boundary. Weeds on wasteland or old manure heaps around the farm should be prevented from seeding. In some cases, this may have to be balanced with biodiversity needs, such as encouraging beneficial insects on particular flowering weeds.

Livestock Feeding/Watering

Water and feeding points must be rotated to ensure that excessive dunging does not occur. Perennial weeds like thistle and docks thrive in high-nutrient areas. If livestock have been grazing on these plants, viable seeds can pass through their digestive systems and germinate when conditions are favourable.

TILLAGE

Tillage is the most effective way to reduce the size of the weed seed-bank. Seeds are encouraged to germinate and then cultivated, usually by mechanical means, to kill off the weed seedlings. There are several stages of cultivation – primary, secondary and tertiary.

Primary Tillage

Most farming systems in the UK are tillage based, that is they operate on some method of turning or cultivating the soil. Traditionally, the first implement to be used in cultivation is the plough and it is the first tool to think of in terms of weed control. Ploughing can be very useful in managing weeds. For example, if weed seeds have been allowed to shed on the soil surface, then ploughing – the action of inverting the soil surface-layer to a depth further down the soil profile – will bury the weed seeds to a depth below which many are unable to germinate. This is particularly useful for small-seeded species that cannot germinate successfully from any great depth or for annual grass weeds that have short longevity and may perish at depth. However, it may only be a short-term solution that temporarily buries the problem. Many weed species have seeds that are extremely long-lived, so if the land is ploughed to the same depth the following year, viable seeds may be re-inverted and will germinate if conditions are suitable.

Secondary and Tertiary Tillage

Secondary tillage usually involves the preparation of a fine, flat seedbed into which the crop can be drilled or planted. This is typically achieved by disking or harrowing to a depth of around 10cm. Several aspects of this operation can be managed in terms of weed control and the operations may be repeated one or more times. Tertiary tillage operations are the direct cultivation techniques employed specifically to remove weeds and are dealt with in more detail in the following section on direct weed-control methods.

Timing

If weather and rotation design allow, a 'stale' or 'false' seedbed can be prepared. This involves cultivating the soil around four weeks before drilling to stimulate a flush of weeds from the soil. This can be effective to encourage the first, and usually biggest, flush of weeds from the seedbed. They can be flamed or very gently cultivated to remove them immediately prior to drilling. A tight cropping schedule may not allow for this break or weather conditions may be unsuitable. Seeds require moisture to germinate,

Comb-harrowing beans

particularly to achieve high levels of germination, that is to say a large flush. If the land is prepared early and the weather is very dry, then the positive impact of early cultivation may turn negative, as the land may be at risk from erosion and also lose what little moisture was available. If irrigation is available it could be used to help 'flush out' the weeds.

The timing of seedbed cultivations can also be performed to avoid the peak flush of weeds that may be problematic in a particular crop: for example, avoidance of the main spring-germinating weeds, such as mayweed or fat hen, by drilling maincrop carrots at the beginning of June.

Method of Preparation

It may also be possible to tailor cultivating implements to the weed flora and management, since weed seeds move differently according to the choice of implements used. Studies have shown that the spring tine, for example, tends to move seeds upwards in the soil profile, whilst the rotovator has the net effect of moving seeds further down the soil profile. If you were preparing a stale seedbed, then a spring tine could be useful to retain freshly shed seeds, which you don't wish to become incorporated into the soil, close to the soil surface, from where they can successfully germinate and be 'flushed out' of the weed seed-bank.

Cultivation in Darkness

This technique has received considerable research and press attention. Some seeds require a flash of light to trigger germination. If it is not provided they will remain dormant. Cultivation stimulates emergence by turning the soil and exposing buried seeds to light and, hence, encouraging germination even if the seeds are re-buried. It would, therefore, seem sensible for cultivation to be carried out at night or in darkness to result in less germination. A number of trials have been carried out at night and with large blackout cowls covering the cultivating equipment to assess the impact on weed emergence. Results have been very variable, but as much as a 70 per cent reduction in weed emergence has been reported in some trials.

This variability in results may be due to the weed flora, the weather, the location, soil conditions, the method of cultivation and various interactions of these factors. For example, not all seeds require light to germinate, so the technique will not be effective on all species. Some species require very low light-levels, so moonlight might be enough to trigger their emergence. Light may also reach shallowly buried seeds by filtering through the uppermost soil particles and small cracks. The effect of light on germination will interact with other environmental factors, such as soil moisture and fertility levels and also soil temperature. The size of the seed may influence the impact of light, since small-seeded weeds tend to be more light-responsive than larger seeds. Whilst results are variable, it does appear that dark cultivation may at least delay weed emergence, so, in crops where weed control is critical, the practice would widen the window of opportunity to carry out future weeding operations.

DIRECT CONTROLS

Although the methods described above provide a good basis for weed-management programmes, it is likely that at least some weeds will still establish in vegetable crops. At this point some more direct action against these weeds is normally taken, to prevent yield loss in the crop. However, any emerging weed flora should be viewed in the context of the damage it is likely to do, the cost of taking action and the long-term costs of not taking any action (for instance, adding to the weed seed-bank). These factors have been posed as a series of questions by Kropff *et al.* (in 'Proceedings of the 13th International IFOAM Scientific Conference in Basel', 28–31 August 2000) and we have taken these as an aid to deciding whether or not to carry out any direct weed-management operation. Briefly, growers should ask: is weed control needed? If so, when is control needed? And where is control needed? Finally, having established that action needs to be taken, which method of control should be used?

Is Weed Control Needed?

Clearly, a combination of many factors will influence the decision whether or not to weed. A good working knowledge of the weed flora and the crop – and, ideally, previous weed-management experience on the particular site – will be indispensable. Applying an understanding of the biology of the weeds on a specific site (life cycle, growth habit, potential reproductive capacity) will help to deal with them appropriately. It will be particularly important to consider whether the main weeds are annual, biennial or perennial, as this will influence the choice of appropriate methods to use and the weeds' ability to resist any management programme.

A weed can be defined as a plant growing where it is not wanted, so the definition of a weed depends on the individual. 'Weeds' will compete directly with a sown or planted crop for space, light, water and nutrients. To prevent yield losses, the weeds need to be removed when they are actively in competition with the crop. Without some form of weed control marketable yield can be devastated and reductions in fresh weight of as much as 96 per cent have been reported in trials with salad onions. There are a number of other reasons why weeds may need to be removed. Their physical presence may impair the harvesting process and shed seeds may affect quality of yield (for example, seeds falling into lettuce heads or contamination of crops such as peas). There are often very specific growth and size requirements for the marketing of many field vegetables and these may be seriously compromised by weed competition. Weeds may also encourage crop pests or pathogens and lack of appropriate control may result in a build-up of the weed seed bank. In grassland, whilst some weeds may be a useful source of trace elements, self-medication or extending the grazing season, many suppress grass yield to a significant degree.

Numerous competition studies that have measured the crop yield loss from weed competition have demonstrated that the cost of complete weed removal can financially outweigh the yield benefits. For this reason many studies have attempted to classify weeds according to their relative competitiveness. For example, a crop may be able to tolerate twenty field pansy plants per square metre, but only one fat hen plant. One way of achieving this classification in an objective and practical way is through the use of economic thresholds. These thresholds have been calculated on the density of a given weed species at which the cost of control would just equal the financial benefits to the crop. Most of these thresholds have, to date, been based on conventional systems where herbicides are the main method of weed removal. It would, therefore, be reasonable to assume that some degree of recalibration of these thresholds would be required to take into account the different costs associated with non-chemical control. In addition, these thresholds do not generally account for the potential added cost of re-infestation – and hence build-up of the weed population – and associated costs that may accumulate over a number of seasons.

Weed management should also be considered in the context of the whole-farm rotation so that weeds, which do not necessarily pose a risk in the immediate crop, may have an impact in later crops if allowed to set. This would also affect perennial crops that will be in the ground for more than one season, because competition in the early stages of growth may affect crop productivity in the longer term. This long-term thinking is critical to organic systems.

When Is Control Needed?

Once the decision to weed has been made, the next question to ask is when should it be done? If weeds are left uncontrolled for too long then they may start to compete with the crop. However, if weeding is made prematurely, the main flush of weeds may not have ceased and so the control operation may need to be repeated. Research on field vegetables has shown that, in many cases, a carefully timed, short, weed-free period (sometimes called the 'critical period') or a single weeding, may be all that is required to prevent yield loss. In one example, in bulb onions, there were no adverse effects on yield from weeds up to five weeks after 50 per cent of crop emergence. However, weeds that remained after week five were shown to reduce yield by 4 per cent for every day they were left uncontrolled over the following two weeks. This two-week period defined the critical weed-free period for that particular cropping situation. In contrast, for more competitive crops such as swede, it has been shown in trials that one single weed-removal operation (as opposed to a weed-free or critical period) approximately six weeks after crop sowing was all that was needed to maintain yields equivalent to that of a crop maintained weed-free throughout the season. The optimum weeding times for different crops are not fixed; they vary not only with the crop but also the weed population and their relative times of emergence. Generally, the earlier the emergence of the weeds compared with the crop, the more competitive the weeds are likely to be. Weeds that emerge late tend to exert little competitive effect – but can have an impact on seed return to the weed seed bank and harvesting.

The optimum timing of weed removal is largely dependent on the prevailing environmental conditions combined with the inherent biology and relative emergence dynamics of the crop and weed species involved. However, there are some factors that can be manipulated by the grower to give the crop a competitive advantage. Such techniques may include, for example, the use of module-raised transplants or of stale seedbeds to put the weeds at a disadvantage. These practices tend to widen the 'weeding window' (a period during which weeds need only be removed once) and hence they make the absolute timing of weed removal less critical. Widening the weeding window is desirable, since it reduces pressure on the farmer to make a weeding operation within a sometimes very narrow time-period, which may be difficult because of weather or other factors.

Taken together, the factors involved in answering the questions whether and when to weed indicate that the weeding of each crop in the rotation needs be planned and there needs to be some element of flexibility in the planning process. The scheduling of weeding operations in different crops in one season will need to be coordinated, although the exact timing will depend on several factors – most importantly the type of crop and weed in question, followed by issues such as such as soil conditions, weather conditions and time of the year.

Where Is Control Needed?

If a weed-control operation is required, it may not need to be carried out evenly across a whole field and, in fact, it is often possible, or even necessary, to target the action in specific areas. The type of action taken might change as the season progresses and as the crop and weed flora develops. There are four specific areas in which weeding operations are often targeted:

- broad-spectrum – weeding across the entire area
- inter-row – weed machinery is focused between the crop rows
- intra-row – weeding is carried out in the crop row itself
- in patches – specific patches are targeted by hand or machine.

Crop inspections, looking at where the weeds are developing, combined with previous experience are likely to indicate where control is best targeted.

Generally, the first weeding operation of a season is usually carried out across the entire cropped area. At this early stage, crop establishment will benefit most from a weed-free environment, to minimize or prevent competition particularly in, and directly alongside, the crop row. If the crop is particularly vigorous and able to form a continuous canopy, it is likely that subsequent intra-row weeding will not be needed. It may be possible in some situations just to focus on weed patches, in cases where there are problematic local infestations. As the season progresses, a broad-spectrum weeding approach may be most effective for some crops, particularly in narrowly drilled crops (like cereals), where harrows are used across the whole soil surface. Harrowing can be performed blind before the crop has emerged and then at desired intervals in the season. Harrowing is also being used more extensively in row crops, particularly in transplanted or strongly rooted crops that can withstand the uprooting action of the tines. This ensures an even working of the bed and it can be very effective when used frequently at the white thread (that is, just germinated below the soil surface) or cotyledon stage. Other crops will demand other techniques and these are discussed more fully in the next section.

Which Method of Control?

This will obviously depend upon the answer to the above questions and especially upon where the control is most needed. There are two main categories of control: physical and biological. In practice, physical control forms the majority of post-sowing methods used commercially today.

Physical Control Methods

Physical controls can be divided into various categories, including mechanical (broad-spectrum, inter-row or intra-row), manual (removal by hand or hoeing), thermal (flaming or infra-red) and mulching (physically excluding light from reaching the soil surface). Physical control will therefore depend on matching the weeding implement with the weed flora to be managed.

MECHANICAL

Mechanical weed-management is often the first method that comes to mind when people are asked how organic farmers control weeds. However, it is clear that there are many factors to consider before a farmer takes direct action with machinery and organic weed-management is somewhat more complex than replacing a chemical with a mechanical application. This form of tillage will kill weeds by burying, cutting or uprooting. Tools without a cutting action are only effective on small weeds. The weather and soil conditions under which the operation is carried out will have a major influence on the efficacy. Some advisors will suggest that if weeds have emerged you are already late with your weeding operations. It is suggested that if weeds are in the white thread stage then this is the best time to kill them.

Broad-Spectrum: broad-spectrum control will aim to weed the whole bed or field. Examples of this type of machinery include chain- or drag-harrows and spring- or flexi-tines. Harrows are more rigid in construction, with steel spikes. They stir the soil to a depth of 20–40mm (8–16in) and are most effective on weeds at smaller growth stages (up to 20mm (8in) height). Flexi-tines work in a similar way but are formed from a coiled loop of metal or are spring-mounted – flat or round tines are available. They vibrate through the soil and glide around obstructions. This vibrating action can, again, uproot small-sized weeds. The harrows or tines are

A typical tine-weeder

located on flexible mountings and the angle of teeth or tines is adjustable, depending upon the degree of attack or aggression required.

Inter-Row: as the name suggests, these implements work between the crop rows. The actual adjustment of the implement can be more important than the choice of equipment, as a multitude of machines are available. Some are powered, some not; others are ground driven. Some machines are front-mounted, whilst others are rear-mounted or carried on specific tool-bars. Many of the rear-mounted machines may also require a second operator, to steer as close to the crop rows as possible. A quick resume of the main types is detailed below:

Non-powered sweeps – these include shares, ducksfeet and hoes. The main methods of kill for these types of machines are by cutting and some burial action. There are various different types of blade and mounting methods available. Hoes are typically 'A'- or 'L'-shaped when viewed in plan. The working depth is around 20–40mm (8–16in).

Non-powered rotary cultivators – these are machines like the basket weeder (or cage weeder). The weeder uproots or strips leaves from weeds. The machine has two horizontal axles on which the rotating baskets are mounted. These axles are connected via a chain-and-sprocket arrangement, causing rotation at slightly different speeds, the first driving the second. Telescoping

Steerage hoe in leeks

baskets allow adjustment of the row spacing. The baskets uproot weeds in the upper 25mm (1in) of soil. They are particularly effective when used in conjunction with an inter-row hoe that breaks the soil surface beforehand. They are effective where there are big stones or long-stemmed residues. Another example of this type of machine is the star hoe or rotary harrow. It is ground driven but has an aggressive weeding action. Discs can be adjusted to move soil away from or into the crop row. The working depth is around 50mm at 8–12km/hr (2in at 5–8mph). One machine, the split hoe, incorporates two designs: a cutting movement with hoe blades and a brushing movement that uncovers uprooted weeds. It can be used on bigger weeds and in moisture conditions than the standard hoe design.

Powered rotary cultivators – these machines use cutting, uprooting and burial to kill the weeds. They are powered and comprise 'L'-shaped blades fitted onto a horizontal axle that is usually covered. They have an aggressive weeding action, pulverizing the soil, incorporating weeds and mixing the soil to a depth of around 120mm (5in). This intensive cultivation removes larger weeds. Similarly, brush weeders have rotating polypropylene brushes that uproot, bury and strip weeds. The most common type rotates on a horizontal axle and can be fixed at the front of the tractor or rear-mounted with a second operator. The brushes are driven, so that the intensity of the work is determined by the speed of rotation and forward

Brush-weeding carrots

direction of the machine. Protective tunnels, available in varying widths, cover the crop row. Brush weeders work the upper soil to a depth of around 40mm (2in) and perform better in friable soil-conditions. They can work at higher soil-moisture conditions than hoes.

Precision guidance weeding systems – are now being developed which allow machinery to be driven even closer to the crop rows. These involve some kind of computer-actuated guidance system, which can differentiate between crop and weed. Several models, which have been attached to hoes, are now available in the UK. The limitation is still that the intra-row area is not weeded. Other machines are now being developed which can move into the crop rows and selectively remove the weeds.

Finger weeder being used in transplanted brassicas

Intra-Row: removal of weeds in the crop row is the most difficult and delicate area in which to control weeds by mechanical methods, due to the higher potential for harming the crop. Some level of intra-row weeding can be obtained with well-set-up inter-row machines, such as the brush weeder or the rotary harrow. These will move soil into the crop row, covering and smothering the small weeds between the crop plants. There are, however, a growing number of machines available that are capable of intra-row weeding.

Finger weeders – these operate with two rubber discs of finger-like projections, which are angled down and into the crop, one mounted either side of the row. These are attached to metal ground-driven spikes. These fingers work the intra-row soil, interlocking and moving between the crop plants. The crop needs to be well-established to avoid the uprooting effect, so it is essential that there is a difference in size between the crop and weeds. The distance between the discs can be altered to allow a gentle or more aggressive weeding action. The machine works best on loose soils and is often mounted on a tool bar behind a set of hoes to break the soil crust.

Torsion weeders – these have a simple design comprising two spring-tines angled backwards and downwards either side of the crop row. The coiled base allows the tips to flex with the soil contours around established plants, uprooting small weeds (those in the white-thread stage) in the row.

The position of the tines can be altered, depending on the level of aggression required.

MANUAL

Manual methods of weed control are still widely used in organic systems. It is important to perform an effective first weed, to prevent weed competition in intensive horticultural-cropping. This will often include following a mechanical inter-row weeding operation with manual labour, to thoroughly remove weeds in the crop row. This can be by means of groups of workers hand-weeding crops or hoeing using hand-held implements. Sometimes these operations can be made more effective or efficient with the aid of machinery, for example by using teams of workers lying on a flat-bed weeder pulled by a tractor. This can work well where a flat-bed weeder follows an inter-row machine, so the minimum amount of hand weeding is performed and only where the inter-row weeder has not reached.

There are some more-mechanized hand-held kits available such as the push hoe, which incorporates a wheel to ease the hoeing process, though this is not as flexible as it cannot cover the crop-row area. Often farm sales are good places to locate useful weeding-equipment, some of which was designed prior to the herbicide revolution, but can still be very useful in modern systems. An alternative is to buy hand-held equipment abroad,

A flat-bed weeder in operation

where hand labour is more widely used and greater attention has been paid to implement design.

Hand rogueing or pulling is also a widely used technique for patches of weeds or removal of difficult-to-control species, such as docks. Hand-held tools have been developed in the UK to remove specific weeds, for example for ragwort, creeping thistle or docks. Removal of individual plants is best achieved with a gang of people, say several people moving methodically through a field pulling and then carting particular weeds off the land. More work on the efficacy and benefits of hand labour throughout the cropping rotation is required.

THERMAL

Two main methods of thermal weed control are currently being used commercially: flame and infra-red weeding. Other methods are also being researched but are still at the trial stage. Both flame and infra-red weeders involve combustion of liquefied petroleum gas. Flame weeding targets a flame directly at the soil surface. Infra-red burners use the flame to heat a ceramic or metal surface that then radiates heat towards the ground. The method of kill is the same in both machines. The plant cell contents are boiled, causing the cell walls to burst, thus desiccating the plant. The temperature needs to reach above 100°C for less than one second in order to be effective.

This method of weed control has many applications. It can be used to produce a stale seedbed, for pre-emergence weed-control in slow-germinating crops, such as carrots, in post-emergence broad-spectrum weed-control in tolerant crops, such as onions, and as a selective post-emergence treatment between crop rows or in patches. A range of different machines is available. The final choice will depend on the size of the operation, the type of cropping and the capital available to spend on machinery.

MULCHING

This provides a physical barrier over the soil surface through which light cannot penetrate. It prevents weed-seed germination and suppresses emerged seedlings. There are three main types:

- sheeted – a layer of material such as plastic, paper or woven fabric covers the soil surface
- particle – a layer of material such as straw, bark or cut residues from the previous crop is spread over the soil surface
- living – a low-growing ground-covering crop is sown.

Crops can be planted or sown through these mulches. Mulches can be useful in more than one area of organic system. For example, some pests have been shown to be confused by an entire ground-cover of green and will land on the mulch as well as the crop, resulting in less crop damage. Living mulches, such as clover under a cereal crop, will improve soil structure and conserve soil nutrients and moisture (*see* Chapter 4).

Biological Control Methods

Biological weed-control involves the release of natural enemies and pathogens for biocontrol of target weeds. In general terms, biological control has been categorized into three different forms:

- classical biocontrol involves the release of exotic natural enemies to control exotic weeds
- inundative control involves the mass production and release of native natural enemies against native weeds, for example rust fungus
- conservative control is an indirect method that manipulates the farm habitat to, for instance, increase the population of insects feeding on the weed or reduce the natural level of pests that attack the native insects which feed on the target native weed.

The aim of biological control is to shift the balance of competitive weed–crop interactions in favour of the crop. In other words, the released or encouraged rust, fungus or insect may not necessarily kill the target weed but can reduce its vigour and competitive ability. Although this has traditionally been a popular area of research, there are, as yet, few practical applications. The successes recorded have generally been seen for the first type of biological control on weeds, which themselves have been introduced into a new area, and where their natural herbivores are absent until deliberately introduced in a biological-control programme (for example, biological control of thistles in Australia).

Allelopathy

Allelopathy refers to the direct or indirect chemical-effects of one plant on the germination, growth or development of neighbouring plants. It is usually expressed through the release of chemicals while the plant is growing or from plant residues after it dies. They can be released from around the germinating seed, in exudates from plant roots, in leachates from the aerial part of the plant and in volatile emissions from the growing plant. Both crops and weeds are capable of producing these compounds, which have presumably evolved, at least in part, to enable plants to reduce competition from nearby plants. From this idea it follows that allelopathy could be useful to organic growers in suppressing weeds.

Allelopathy could be used to manipulate the crop/weed balance by increasing the toxicity of the crop plants to weeds or reducing weed germination in the direct area of the crop – the most difficult area to control weeds physically. Alternatively, a mulched residue of an allelopathic cover-crop could prevent weed germination and allow a transplanted crop to be grown, producing a residual weed-control effect. These allelochemicals could be isolated and form the basis of 'natural' herbicides that, in the long run, might prove acceptable under organic standards.

There is ongoing research into the so-called allelochemical properties of plants, particularly in laboratory investigations, to ascertain which species have these abilities. There has been little practical field research and the level of impact of allelopathy in temperate systems is not known, but is thought to be worthy of further investigation.

Chapter 6

Pest and Disease Management

Pests and diseases can cause serious losses to organic growers, both in crop yield and monetary returns. Organic pest- and disease-management programmes should be designed according to organic principles. They should, therefore, be planned so as to maximize the probability that pest and disease problems will be kept in check without excessive intervention on the part of the grower. In practice, this means that growers should work towards reducing pests or disease to levels below that at which they cause economic losses, using a range of preventative techniques. They should use curative measures only as a final resort. Total elimination of pests and diseases is impossible in practice and, even if it could be achieved within the limits set by organic principles, would anyway incur the risk of a resurgence of pests and diseases, as natural enemies and competitors to plant pathogens are lost from the production system.

This chapter develops the concepts and principles behind designing pest- and disease-management programmes for organic-vegetable systems. It discusses some of the common techniques available to organic growers that might help to prevent pest and disease problems, before going on to consider some possible curative approaches that might be necessary when faced with serious crop losses. Finally, we consider combining the principles and practice to formulate pest- and disease-management strategies for vegetable production.

In the text, and for brevity when discussing general principles, we have used the term 'pathogen' or 'plant pathogen' to denote pests and diseases in general. However, the term 'pests and diseases' covers a wide range of organisms that attack crop plants. Diseases include viruses, bacteria and fungi. Pests include nematodes, insects, molluscs and vertebrate pests, such as birds and mice. Obviously, it would be difficult to include detailed information about all these in this text. In this chapter the common pests and diseases likely to be encountered in vegetable production are mentioned in the context of management techniques by way of illustration. Further details of pests and diseases of specific crops are given in the relevant parts of Chapters 3 and 12, which discuss production of vegetable crops in more detail. Additional details can also be found in many of the numerous specialist texts that have been published on pests and diseases of vegetables (*see* Further Reading).

PEST- AND DISEASE-MANAGEMENT CONCEPTS FOR ORGANIC PRODUCTION

At their heart, pest- and disease-management programmes in organic vegetable-production systems aim to prevent economically damaging levels of pests and diseases building up on crop plants. Management programmes need to be designed within the framework of organic principles and any practices used will have to be allowable under organic standards.

In general, and in keeping with the holistic approach to organic farm-management, organic growers should work towards producing well-nourished crop plants within a biologically diverse farm environment. Promoting a healthy soil, with good structure, high biological activity and containing sufficient nutrients, will promote the growth of vigorous crop plants that are, in turn, unattractive to or resistant to pests and diseases. In addition to this, a diverse ecological farm production system should maintain a dynamic that favours the crop over pests and diseases by reducing the area of monoculture or suitable habitat available to crop pathogens, by increasing the presence of competitors or natural enemies of the crop pathogens and by decreasing the probability of transfer of pathogens between susceptible crops.

PEST- AND DISEASE-MANAGEMENT PRACTICES

These basic concepts have been translated into a set of practices that organic growers can use to manage crop pathogens in their farm systems. For convenience, these can be divided into two approaches: strategic and tactical. Strategic practices are techniques that are primarily aimed at preventing pest and disease problems arising in the first place and generally take the whole farm system into account. In taking strategic measures there is a range of basic practices that can be considered central to all pest and disease management on organic farms. These include maintaining and promoting soil fertility, using crop rotations, using crop varieties resistant to pests and diseases, promoting biodiversity and promoting and taking basic sanitary precautions to prevent the spread of pathogens. Tactical practices can be curative or preventative and are normally aimed

Slugs are possibly the most troublesome pest of field vegetables

at specific pest, disease and crop combinations, usually where serious crop loss is anticipated, and will often be elements within a rotation or field. Once again, there is a range of cultural, biological and chemical techniques that can be employed. There is, necessarily, a great deal of overlap in these approaches and techniques and within an organic farming-system the approaches should be harmonized to work with each other as far as is practicably possible.

Many of these practices and techniques were, or have become again, common practice in conventional farm systems. Many have also become subject to organic standards and their use has been codified. This particularly applies to more disruptive practices, such as using chemical techniques where many practices are restricted and permission must be obtained from a certifying body before employing them. Other practices, such as length of time between susceptible crops in rotations, may also be defined and the organic standards of certifying bodies should be used as a guide when formulating pest- and disease-management strategies.

Strategic Management: Basic Organic Management Practices

Central to all organic pest- and disease-management practices are strategic practices aimed at preventing problems arising. Two practices widely used to achieve this are crop rotation and use of resistant crop varieties. In addition to these, organic farmers should also maintain soil fertility, promote biodiversity in and around crops and take basic sanitary precautions to prevent the spread of pathogens between fields and between farms. The implications for pest and disease management, the techniques used and the effects they can have are discussed below for each of these cases.

Soil Fertility

The general importance of soil fertility, chemical, physical and biological, has been stressed and discussed at length in Chapter 4. Generally, a fertile soil is one that contains sufficient available nutrients to allow plant development, a good physical structure that allows development of healthy roots and high levels of biological activity to allow efficient cycling of nutrients. This is important in pest and disease management for a number of reasons, both direct and indirect.

Indirectly, good growing conditions affect the plant's ability to defend itself. Plants have a well-developed defensive system that is more likely to function effectively when they are adequately nourished. A soil that promotes vigorous plant growth will, therefore, also help the plant resist pests and diseases in this respect. Plant defensive-systems (*see* below) rely on the production of secondary metabolites, which directly or indirectly affect herbivores or diseases and which, in general, are more likely to be effectively produced in plants under good growing conditions where the plant can obtain an adequate supply of macro- and micro-nutrients. Pests in particular are often attracted to plants that are stressed, which can be an indication that the plant is not able to defend itself in some way. It is also well known that some groups of insects, such as aphids, are attracted to plants that contain high levels of soluble nitrogen, so that a balanced supply of nutrients, such as that obtained from a healthy organic soil, is beneficial.

More directly, a vigorous soil ecology can directly suppress soil-borne pathogens, including nematodes and insects. Plant disease pathogenic organisms often only compete weakly with other soil organisms and so can be suppressed in biologically active soils. Their resting spores or other structures can also be directly attacked and destroyed by predatory soil organisms. The phenomenon of soils suppressive to disease is well documented and, although the exact mechanisms are often unclear, it is usually attributable to the presence of organisms in the soil that directly attack the disease-causing organism.

Rotation

Crop rotations are integral to organic-farming practice for many reasons, one of which is pest and disease management (*see* also Chapter 7). In many cases pests and diseases are unable to survive in the absence of the host crop and so absence of the host will cause their numbers to decline. The mechanisms at work are varied and can include a simple 'starvation' effect, but can also include their being outcompeted by other soil organisms or their direct destruction by predatory organisms during the rotation. In such cases, rotations should allow for the longest possible period between growing crops of the same family. Adding cover crops and green manures can also have indirect beneficial effects (through improved soil fertility and plant nourishment as discussed above) and sometimes can have a direct biofumigant effect, cleaning the soil of susceptible pathogens. For instance, when they decompose in soil, brassicas release various compounds that affect the populations of micro-organisms and there are some situations in which they can diminish populations of plant pathogens in the soil.

Disease management by rotation is most useful against soil-borne pests and diseases such as cyst nematodes in potatoes, white rot in alliums and clubroot in brassicas, although some of these might persist as resting cysts or spores in the soil for long periods. In any case, longer and more diverse rotations will, in general, be better for disease control. Although rotation will not generally be effective where pests or diseases are spread by air, some pests like the carrot root fly are weak fliers and do not generally disperse large distances. In this case, ensuring that fields of carrots in a rotation are placed some distance apart can help to reduce carrot fly attack. It should be borne in mind that, in some cases, rotations can work to promote problems, for example wireworms (larvae of click beetles) can be a problem in vegetable crops such as potatoes grown immediately after grass-clover leys.

Resistant Varieties

Plant varieties differ in their susceptibility to pathogens. Plant resistance is expressed in many different ways and is often termed qualitative, where there is no visible effect of the pathogen, or quantitative, where the pathogen is visible but developing only slowly. Resistance to disease ranges from hypersensitive immune-reactions in which the pathogen is isolated in dead plant-cells (often seen as small, black, necrotic flecks) to the production of sophisticated chemical defences that hinder or kill the pathogen. Resistance mechanisms to disease can also include morphological characteristics, such as thick waxy layers, that prevent pathogens gaining access to plant tissue. In the case of herbivorous pests, such as insects and nematodes, plants can produce a range of chemicals and morphological structures (like hairs) that deter pests from feeding or actually impair their development. Some plants can even produce semio-chemicals that signal the plant is under attack and which can serve to attract parasitoids of insect herbivores or chemicals that mimic vertebrate hormones, such as oestrogen, that can modify vertebrate behaviour. In some cases plants are tolerant to pathogen attack and can produce compensatory growth.

In more practical terms, resistance is referred to as specific (vertical) or non-specific (horizontal). Specific resistance relies on plant breeders matching specific virulent genes in the pathogen with resistant genes in the plant and this means that the variety is resistant to specific races of the pathogen. This is the case with downy-mildew-resistant lettuce, which is often quoted as having specific race-resistance (e.g. BL 1–7, 10–16 indicates that a variety is resistant to the disease races specified by the numbers). In contrast, non-specific resistance is effective against all strains of a pathogen and is usually associated with a number of genes or gene complexes rather than single genes. Specific resistance is highly effective even under high disease-pressure, but because it strongly selects strains of pathogen that can overcome the resistance, it often breaks down over a number of seasons, as in the case of downy mildew in lettuce. In contrast, although non-specific resistance can break down in high disease-pressure years or when environmental conditions give an edge to the pathogen, it is generally acceptable in normal cropping conditions and can be expected to be long-term. Non-specific resistance is often thought of as more appropriate in organic production systems but is, unfortunately, harder to define and select in breeding programmes. Indeed, some people argue that such broad resistance is the normal condition for plants, as they are obviously generally resistant to most plant diseases. In this case it becomes a question of breeding or selecting 'healthy plants', which, in the end, is a subjective judgement.

There is a limited amount of information available on the pest and disease resistance of varieties grown under organic conditions. However, in most cases, resistance ratings from non-organic trials should provide at least a useful comparison between varieties and give an indication of likely performance in this respect. Information on pest and disease resistance is routinely given in seed catalogues produced by seed companies and is also provided in information produced in the national variety-trials system run by the National Institute of Agricultural Botany (NIAB) (*see also* Chapter 3).

Promoting Biodiversity

Biodiversity is the sum of species in a given area and is reflected in the number of species present and the variety of interactions that occur between them. In the case of farms, this obviously includes the genetic diversity of the crop and livestock species present. Generally, the higher

Cabbage root-fly damage in brassicas

the number of species and the greater the number of interactions between them in a given area, then the higher is the biodiversity. Ecological studies indicate that, in general, systems with higher diversity have a higher and more stable productivity, attributes that can also be seen as desirable in agricultural systems. This is partly because resources are used more efficiently, but also because there are usually no unchecked outbreaks of pests and diseases. The mechanisms responsible for reduced numbers of pests and diseases include increased predator or parasite populations, provision of alternative hosts and prey for natural enemies, decreased colonization or reproductive rates of pest organisms, feeding inhibition or repellancy effects, prevention of movement and emigration of pests and diseases, as well as increased synchrony between the life cycles of pests, diseases and their natural enemies. Some of these factors are likely to act in tandem, although they can also work against one another in some cases.

Growers can affect diversity on their holdings at many levels. At the field scale the arrangement of crops, both spatially and temporally, can be used to increase biological and genetic diversity. So, more complex rotations over longer periods, mixtures of varieties in a field and mixed species or intercropping can all serve to increase complexity and diversity to the benefit of pest and disease management. The use of flower strips and beetle banks in fields can also provide refugia for natural enemies of pests and break up areas of monoculture to impede disease spread. The type and intensity of management operations will also affect biodiversity. So reducing the number of tillage operations can help promote biological activity in the soil, as can leaving or burying crop residues. Tolerating at least some weeds can be important for promoting diversity within crops. Sensitive headland management can also work to increase on-farm diversity. At the level of the farm and wider landscape, the management of non-crop vegetation can be important, providing a mosaic of fields, ideally with a perennial crop component and wildlife areas that all help to promote biodiversity, especially of larger animals and birds. Increased biodiversity and increased enterprise diversity are also likely to be linked, so that the complexity in the farm landscape will necessarily follow from or lead to the development of different and diverse enterprises that use the many different resources within the farm area. Such diversity is likely to involve different types of costs and different social structures, including different types of market and business practice. These are discussed in more detail in Chapters 9 and 10.

Hygiene and Sanitation

Hygiene and sanitation can be an important but often overlooked aspect of pest and disease management on-farm. Preventing the introduction or spread of pathogens precludes the need to treat them in the first place. The extent to which this is possible depends on the pathogen and the costs involved. Practices such as ensuring that only clean, certified seed is used on the farm and that transplants are free from pests and diseases are an important first step to establishing a healthy crop. Own saved seed can be checked for pest or disease loading before being sown. Cleanliness in on-farm operations can prevent the spread of pathogens between fields on a farm.

Soil-borne diseases (such as clubroot in brassicas or root-knot nematodes) are often spread on equipment (tractors, boots, implements) and these should be routinely cleaned down after operations in any particular field. Removing crop debris can, in some cases, help prevent spread of disease (such as *Botrytis* rots), as can the removal of volunteer plants (potato blight) or weeds (for instance, dark leaf-spot in brassicas can be found on charlock). Attention should be paid to neighbouring crops (for instance, oilseed rape and brassicas can harbour infestations of cabbage root fly, which can fly to newly planted crops). Physical removal and destruction of infested or diseased plants may be useful in some situations (especially in smaller area or on perennials) but usually only serves to slow progress, especially where plant diseases are concerned.

It may be possible to avoid sowing or growing crops in periods in which the crop is susceptible to pathogens. For instance, even small numbers of carrot flies may cause significant damage to carrots and many growers plant main-season carrots in June in order to avoid the first peak of carrot flies. Sowing in cold, damp conditions can promote the incidence of foot rots in peas when the disease-causing organisms are present, but the disease might be avoided when the soil is warmer and drier. In some cases, and where high and constant losses are experienced, it might be necessary to avoid growing a crop all together.

Tactical Management: Intervention against Pests and Diseases

A large range of practices have been developed that can be used to try to tactically manage pests and diseases in organic farming systems. They cover a large range of cultural practices, direct physical or chemical controls and, increasingly, biological control agents. These are normally, but not always, employed once the pest or disease has been observed and, therefore, are generally more curative in nature. It has also become increasingly possible to receive forecasts that predict likely pest and disease attacks and, in this case, many of the measures can be employed at times of high risk, before the pathogen has become established. In the UK, forecasting systems exist for a range of pests (aphids, cabbage pests, carrot fly) and diseases (lettuce downy mildew, brassica foliar diseases).

Cultural Practices

A large number of cultural practices can be adopted to manage specific pathogen organisms. In some cases they will have secondary or even undesirable effects on other aspects of the production system. In this case, their use will depend on the balance between costs and benefits weighed for the crop or farm situation. Here we have mentioned some techniques that might be useful in organic vegetable-production systems.

Manipulating the time of sowing and the crop density have traditionally been used to manage diseases. Generally, more open canopies (wider spacing between plants and between rows) will reduce spread of plant diseases, as leaves dry more quickly and humidity is reduced in the canopy, conditions which reduce disease establishment on plants. Manipulating planting depth can also help in some cases. Shallower planting of potatoes will help to reduce the incidence of stem canker in spring, as the shoots are not in contact with wet and cold soil (conditions which promote the disease) for as long as they are when tubers are buried deeply. Similarly, shallow sowing of bean seed to promote rapid emergence can help reduce stem and root rots.

Other methods can be used to manipulate the environment around the crop. Obviously, adequate and careful cultivation of soil prior to sowing or planting will allow crops to establish quickly. This might be important for such pests as flea beetles on brassicas, where rapid crop establishment and quick growth is necessary to outgrow pest damage. In this case, irrigation to aid establishment might also be beneficial. Raised beds and mulches are well-known methods that indirectly manipulate pathogen attacks by modifying soil conditions, generally warming it up and drying it out. A large range of mulching materials is available: from living mulches and cut plant materials to plastic sheeting. In some cases, it is also possible that mulches might encourage some types of pests such as slugs, which will shelter under various types of mulching material. Time of weeding can also be important. Weeds, although having detrimental effects on crop yield, can also provide food or alternative prey for parasites or predators of pathogens. They can also help to confuse pests trying to find crop plants or to disguise crop plants against the background. In this respect it might be beneficial to have at least some background weeds at some stage of the crop cycle.

Polycropping, that is using combinations of crops together or in an overlapping sequence, can mitigate the effects of pests and diseases. This functions by reducing the chance of pests or diseases encountering a suitable host in mixtures and/or by confusing pest host-finding behaviour. For instance, the cabbage root flies will often fly away from cabbage host plants if they encounter leaves from companion plants whilst they are assessing the plant as to suitability and locating the stem collar at ground level where they lay their eggs. Intercropping can lead to competition between crops and reduced yields and can also complicate harvesting, but some of the effects might be seen from other arrangements, like strip cropping. Interestingly, manipulating weeding dates can have similar effects to intercropping, under some circumstances, with the obvious danger that the weeds will outcompete the crop if they are left too long. Intercropping regimes are likely to be specific to farm conditions and would generally need to be tested over a number of seasons to weigh the costs and benefits.

Crop covers and barriers are widely used in organic vegetable farming (*see also* Chapter 8). Covers, such as fleeces or meshes laid over crops, serve to prevent pests entering a crop. They can be effective in crops such as brassicas against pests like cabbage white caterpillars, diamondback moth and mealy aphid and are widely used in carrots to control carrot fly damage. Covers and fleeces can be labour intensive, as they need to be removed for weeding and other farm operations. They also modify the climate under the canopy so as to make it warmer and more humid, which can also work to promote diseases or, if pests like aphids gain access, allow them to multiply unchecked under a protective cover. Barriers are normally used to prevent larger mammals, such as rabbits, badgers or deer, gaining access to and feeding on crops. Cages can exclude birds, but this would not normally be practicable in vegetable production. Smaller barriers may be used in specific circumstances, such as felt collars around individual brassica plants to prevent cabbage root flies laying eggs in the soil, but would not normally be practicable to use on a large scale.

Scaring devices are also commonly used against birds and mammals. Some devices produce loud noises at random to scare off predators and others work by creating alarming visual stimuli. Others hum in the wind or flutter and cause a rustling noise. Scarecrows can also be used to create the impression that people are present in a field. However, birds (and mammals) generally become quickly accustomed to such devices and, especially when no alternative feeding sites are available, will quickly

Crop covers can help to exclude pests

learn to ignore them. The best devices generally work to trigger innate behaviours such as predator avoidance. Kites with bird of prey silhouettes, for example, are often used to scare pigeons and some seem to be generally more effective than other devices.

As an ultimate control method, pests and diseases can be physically removed. This method is only likely to be effective if carried out regularly and rigorously. Traps have traditionally been used to locally control moles, mice, rats, rabbits and some birds, although such traps have become socially less acceptable in recent years. Water traps and sticky traps can also be used against insect pests. These rely on colours or pheromones to lure insects onto either a sticky card or water-filled containers and can be used against thrips, aphids and moths, amongst other pests. Slugs are also often lured into water-filled or bran-baited traps. A device that causes flea beetles to jump off plants by knocking the plants has also been promoted, but only works in conjunction with a device that collects the beetles, for example on a grease-coated plate. Similarly, tractor-mounted collecting devices have been used against Colorado beetle (a potato pest) on the European continent. In order to have an appreciable pest-control effect, a significant and usually impractically large proportion of the population needs to be caught; so traps are more normally used to monitor pest and disease populations, as an aid to making decisions on taking further control measures.

Manual removal is also a method of removing pests. However, this is probably only practicable in small areas with larger-sized pest insects, such as caterpillars on plants or cutworms in soil. The removal of diseased tissue on plants will only usually be a stopgap measure that will hinder but not prevent spread of disease.

Biological Controls

Biological control involves the manipulation or release of natural enemies and pathogens in order to control plant pathogens. The aim of biological control programmes is to shift the balance of competition between the pathogen and the crop in favour of the crop. The biological control agent, normally a fungus or insect, will normally kill the pathogen or at least reduce its vigour and competitive ability. This subject area has increasingly received considerable research attention and some success has been achieved. In practice, there are three categories of biological control:

- classical biological control involves the release of exotic natural enemies to control exotic pests
- inundative control involves the mass production and release of natural enemies

Carrot-fly traps are useful for monitoring populations of carrot fly

- conservation control is an indirect method that manipulates the habitat around or within the crop with the aim of encouraging those organisms that attack the pest or disease.

Classical biological control, that is releasing natural enemies of an introduced crop-pest, has been successfully used against pests that have been introduced into new areas, normally along with a crop from its area of origin. Such programmes are now large and logistically complicated, from the prospecting phase of looking for likely natural enemies in the area of origin, to the quarantine and evaluation phase in which their likely impact in a new area is evaluated, and the final introduction and release phase. We know of no such successful programmes in UK agriculture against pests or diseases, although there have been successes on other continents, including a successful introduction of parasites of cassava mealy bug in Africa, which has helped to manage this damaging pest of an important food crop. There have also been some well-publicized negative side effects of such programmes. For instance, cane toads introduced into Australia to manage sugar-cane pests have themselves become a pest that is causing considerable damage to the indigenous fauna in the north of that country.

Inundative biological control has become more popular in recent years, with the development of technologies capable of mass rearing of biological control agents including viruses, bacteria, fungi and insect parasitoids. In effect, the commercial possibilities for using these agents as 'biological sprays' is much more promising, as they normally have to be reapplied at periodic intervals and this also provides a research impetus. One such biological control agent known as Bt (a bacterium, *Bacillus thuringiensis*) has become widely used in organic systems against lepidoptera caterpillar pests, such as cabbage whites in brassicas. A great deal of research has also been undertaken on viral, fungal, insect and nematode parasites of insect pests. Some, like *Verticillium lecanii*, a fungal parasite of aphids, and various aphid parasitoids, like *Aphidius colemani*, have been developed and are routinely used in glasshouses where the potential for confining the biological control agent under more-or-less controlled environmental conditions ensures adequate pest management. Such control programmes are less successful in field crops, as environmental and other conditions are much more variable. Nematode parasites of slugs (*Phasmarhabditis hermaphrodita*) are available commercially and generally work well on small slugs in moist conditions above 5°C. Research has also been undertaken on control of carrot fly using nematode parasites, with more variable results to date. It is also becoming increasingly easy to obtain insect predators, such as ladybirds, that can be released onto crops before pests, such as aphids, become fully established. In all these cases, costs are likely to be high and control less certain for large areas of vegetable production and, therefore, they are probably more suited to smaller production areas at the moment. A large amount of research work is being undertaken on biological control agents for use against plant diseases and a few commercial formulations are becoming available. For instance, the fungi *Trichoderma harzianum* is available commercially as various seed-treatment products in some countries and is a mycoparasite of several damping-off pathogens, including *Pythium*, *Rhizoctonia* and *Fusarium*. The reliability of these products is dependent on environmental conditions and efficacy of these products is generally still low.

Conservation biological control or habitat management has become increasingly promoted as a biological control technique and has previously

Beetle banks provide refuges for predators and parasites over winter and when the crop is not present

been discussed as a strategic management tool for organic growers. In this case, it obviously works more indirectly than the previous two methods by creating conditions for natural enemies of crop pests and diseases to survive and increase. The mechanisms by which it works are to create and supply resources that biological control agents need. These can be physical or biological. Beetle banks create a refuge for carabid beetles in or around fields and provide a shelter where they can avoid intensive field operations or over-winter. They can move back into fields (as much as 60m) from the beetle banks when the crop is present. Such refuges also potentially offer alternative sources of food for biological control-agents, so that predators or parasites of insects may, for instance, find alternative prey on headlands of fields that tide them over in the absence of crops. Flower strips in headlands can also be a source of food for parasitic insects, the adults of which often depend on flower nectar to survive and produce eggs. Weed management can also be important in managing pests. Weeds are often able to provide alternative food-sources or alternative prey for parasites and predators and this should be balanced against their obvious detrimental effects. Providing shelter, such as lacewing boxes, may also help to increase predator populations, especially in smaller areas. It may also be possible to provide nesting boxes or perches for predatory birds, like owls or kestrels, that help to control small mammals. Hedgerows and small ponds also provide resources for larger predators, including frogs and birds that also consume crop pests, such as slugs and insects. A diverse landscape, in itself, can also hinder the movement of pests and diseases between fields, as it exposes them to predators and parasites.

Chemical Control

'Chemical control' is a catch-all phrase, used in this context to cover the application of a diverse range of products, normally in liquid form. They are generally used to modify pest behaviour or, in some cases, to kill pathogens. They exert their effect either directly on the pathogen itself or can work indirectly through stimulating the plant to resist the pathogen's attack. They are normally based on natural plant extracts, compost teas or inorganic salts.

In order to be effective, any chemical needs to be delivered to the place in which it exerts its effect in sufficient concentration at the right time. This means that equipment should be calibrated and application carried out at the appropriate time. This is especially important with plant extracts, as their persistence in the environment is generally very short, for instance pyrethrum, a natural knock-down insecticide extracted from plants, is rapidly degraded in sunlight. It might be necessary to repeatedly apply some products if the weather is adverse: that is, strong sunlight will tend to degrade products or heavy rain will wash simple solutions off leaves.

Materials applied to control pests and diseases are now subject to extensive control under national legislation in all countries. Before applying such products, producers should ensure that they are complying with such laws that, at base, have been enacted to protect the environment, the grower and the consumer. In addition to any national laws, the application of such products must also be acceptable to the certifying body with which the grower is licensed. Many such products are now restricted for use in organic systems or prohibited. Others, such as copper, used in disease control, are being phased out of use in organic systems over a period of time.

Plant extracts cover a wide range of natural pest-control materials, some of which have been used in traditional farming systems. They cover materials that knock down and/or kill pest insects and these including neem, garlic and pyrethrum. Natural pyrethrum is a powerful chemical that can rapidly knock insects down by affecting their nervous systems and can kill them. The modes of action of garlic and neem are not so well understood, but they can certainly inhibit insect development. Garlic has been shown to kill cabbage root fly eggs or young larvae on contact under laboratory conditions, but efficacy in the field is generally much reduced. Other plant extracts are reputed to have anti-feedant or repellent effects on insect pests. So, extracts of alliums (like garlic) or capsicums (like pepper) seem, under some circumstances, to offer some protection against pests feeding. The modes of action are unknown and it is unwise to assume that something that tastes unpleasant to humans can even be tasted by insects or other pests, as their taste sensors react to a range of different, but usually very specific, chemicals.

Another range of materials, which can include plant extracts and also seaweed extracts and compost teas, serve more to stimulate the plants own defence systems or create conditions under which plant pathogens are outcompeted. 'Acquired resistance' is based on priming the complex immune-responses in the plant, so that the plant readily responds to attack by pathogens. Once again, the exact mechanisms of action are variable and in many cases the details are unknown, but much research work has been devoted to looking at the control of plant diseases using these methods. There is some evidence that seaweed extracts can have these effects, but it is often difficult to determine mechanisms, due to the large number of factors involved, not least the presence of plant hormones in seaweed extracts. Compost teas seem to rely on placing a diverse mixture of micro-organisms on the crop plant, which either directly work to exclude plant pathogens or, once again, prime the plant to react to invasion.

A range of simple inorganic chemicals has traditionally been used in organic-farming systems to control mainly diseases. These are normally based on copper (for instance, copper against potato blight) or sulphur (against a range of plant diseases) and can be effective when applied correctly. Bicarbonate of soda has also been used against plant diseases with some success. Some of these chemicals can have significant and detrimental effects in the wider environment and many are being phased out under organic standards. They are likely to be increasingly restricted under national pesticide legislation as well.

There are a range of miscellaneous products that have been used in organic pest and disease management. The use of soft soap against aphid pests is ubiquitous and can be effective as long as the aphids are directly contacted with the solution. A range of biological control-agents, such as Bt, have being increasingly used in pest control and these have been discussed under inundative biological control above.

GROWER STRATEGIES FOR MANAGING PESTS AND DISEASES

In order to effectively manage pests and diseases, growers will normally aim to use a mixture of strategic and tactical methods. It should be remembered that both diversity and adaptation are features of biological systems as they develop. That is, biological systems tend to become more diverse over time as organisms arrive in an ecosystem and adapt to conditions within the ecosystem. In fact, bare earth, the starting point of most crop cycles, is comparatively rare in ecological systems. From this perspective agro-ecosystems are in some sense trying to hold back the build up of diversity and growers will have to learn how to manage this process and where intervention will be most effective, both in promoting organic principles and in running a business. In general, organic crop-protection programmes should be integral to the whole farm system and they should be based on an accurate identification of pest or disease problems. This should be tempered with a realistic assessment of level of risk of economic loss and growers should move towards long-term strategies for 'organic' pest and disease management.

In designing pest- and disease-management programmes growers should follow a few simple guidelines. The first of these is to get to know the pests or diseases that occur in the farm system. Some time spent noting what pests or diseases occur and consulting some of the many texts available on their behaviour is worthwhile. Understanding their behaviour and habits will also enable a realistic assessment of likely losses. Taking notes and using simple counting methods to estimating pest or pathogen density will also lead to more realistic assessments of likely build-up of damage, if it is done in a systematic way over a few seasons. Basic information like this will help accumulate experience and knowledge that will pay off over time. This can be summed up as identifying the key pests, knowing their biology, recognizing the kind of damage they inflict and understanding economic status. Growers should instigate and maintain monitoring programmes for all pests and diseases. These can be simple systems and need not consume a lot of time compared, say, to the resources that might be devoted to controlling a pest that is not actually causing significant losses in all years. Use forecasting systems where these are available.

Once pests or diseases have been recognized, sources or causes of these pests or diseases can also be identified. Understanding the biology of the crop or resource, especially in the context of how it is regulated by the surrounding ecosystem, can be important. Consider, and identify as far as possible, the key environmental factors that impinge (favourably or unfavourably) upon pest and potential pest species in the ecosystem. Identifying sources or causes will often go a long way to suggesting potential control techniques that will work in any given circumstance. Identify risk for site and year (local problems/changes/fields) and weather (winter/season/forecasts). Pests and diseases can be internal (volunteers/weeds/debris/other crops/soil) or external (transplants/seed/neighbour crops/long range).

Finally, growers will need to respond to problems by thinking about long-term strategic approaches to managing pests and diseases, using the many strategic practices discussed in the first part of this chapter. However, growers are also likely to need to be adaptive and flexible in responding to problems as they arise within seasons and this is especially likely to be true while converting from conventional production. Seek the weak links in the armour of the key pest species and direct control-practices as narrowly as possible at these weak links. Consider concepts, methods and materials that, individually and in concert, will help to permanently suppress or restrain pest and potential pest species. Structure the programme so that it will have the flexibility to adjust to change; in other words, avoid rigidity in a programme that cannot be adjusted to variations from field to field, area to area or year to year. Anticipate unforeseen developments, expect setbacks and move with caution. It might be necessary to accept damage in many circumstances. Above all, be constantly aware of the complexity of the resource ecosystem and the changes that can occur within it. In keeping with organic standards, programmes should avoid broad impact on the resource ecosystem. Whenever possible, consider and develop methods that preserve, complement and augment the biotic and physical mortality factors that characterize the ecosystem and attempt to diversify the ecosystem.

Chapter 7

Field Management and Rotations

Organic farming systems depend to a greater or lesser extent on a diverse rotation. In contrast, non-organic systems often repeat cropping over a number of years or follow much simplified rotations. Organic farms depend on a well-planned and well-designed rotation to ensure their biological and financial viability. Key biological factors in organic rotations include soil-fertility management, maintenance of soil structure, weed management, and pest and disease control. These factors are regulated to some extent by organic standards and organic certification bodies generally make recommendations on good rotational practice based on their standards. These factors have been dealt with in more detail in Chapters 4, 5 and 6. In addition to the biological factors, a rotation must also take into account the financial and marketing aspect of farm management and these factors will be developed in Chapters 10 and 11. This chapter aims to synthesize these elements and develop the theme of the whole-farm system of management and rotations.

PRINCIPLES OF GOOD ORGANIC MANAGEMENT

Rotations are the heart of all organic farming systems in all but a few specialized cases. Well designed and implemented rotations have many benefits which are summarized below. Many of these themes are developed at length in separate chapters.

Maintenance of Soil Fertility and Structure

Organic agriculture does not have the quick-fix solution of non-organic mineral fertilizers to make good the shortcomings of poor soil structure or poor nutrient availability to the crop. Due to this, good soil management is of paramount importance in organic farm management. The principles of good soil management are discussed at length in Chapter 4, but revolve around the need to maintain the soil's natural structure in order to allow the free movement of air and water and to enable plant roots to penetrate the soil extensively, thereby accessing the nutrients in the soil. Nutrients are available in various forms: those stored in mineral form, those added to the soil in farmyard manure, green manures and composts, and those made available through the symbiotic relationships and/or activities of various soil micro-organisms.

Weed Management

Weeds are considered to be one of the priority problems by organic growers. A diverse rotation with a range of crops and cultivation methods can help prevent the build-up of weeds in a farm system. It can also reduce the return of weed seeds to the weed seed-bank. Many of the implications of rotation for weed management are discussed in Chapter 5.

Pest and Disease Management

Once again, organic agriculture does not have pesticides to provide a quick-fix solution to pest or disease outbreaks. As discussed in Chapter 6, rotations are often essential in reducing pest and disease carry-over between crops. This is particularly true of soil-borne pests and diseases. In addition to this, there is considerable evidence that a vigorous soil ecology, as promoted under a sound rotation, can help to promote the health of plants and help them to resist the attacks of plant pathogens.

Marketing

A good rotation will help with marketing, ensuring that the farm business is able to extend over a full season or even year. A diverse rotation can also help to tap into different markets and sell a greater proportion of produce, as markets have different quality specifications. Marketing concepts are further developed in Chapter 10.

Spread of Costs and Financial Risk

Apart from aiding marketing, a diverse rotation will also help to spread the costs of producing organic vegetables over the cropping year. It will also help to smooth-out peaks and troughs in labour demand and can ensure a measure of stability that can enable a business to develop and retain staff skills. A diverse cropping-regime can also ensure that a failure in production or marketing in one particular crop will not be a complete disaster to the whole business.

ROTATIONS

Despite the fact that organic certification bodies may require or prohibit certain practices, there is no one prescribed organic rotation. The design of a rotation needs to draw together the different factors or resources available into a sustainable production system, bearing in mind that this system is likely to be a unique combination of agronomic, economic and geographic conditions. This boils down to a balancing act that maximizes the positive outcomes and minimizes the negatives and which is essentially a play-off between various influences within the farming system. It is also a dynamic process, as a farming system develops over time; for instance, crop requirements may change with fashions, pest and disease problems will wax and wane, weed problems will change as the system develops and labour and machinery provision will alter over time. Even changes in weather patterns due to global warming are likely to become more relevant in the future.

Core Rotational Practices

The two main areas in which skilful management comes into play are in the choice of cropping or rotation and in the appropriate use of machinery. Choices made in these two areas can make the difference between 'good' and 'bad' management practice. However, certain core practices are generally recognized by most certification bodies and growers and these are discussed below.

Achieving a balance between fertility building and exploitative crops: generally nutrients must be replaced by including fertility-building crops in the rotation and/or the use of green manures, composts and farm yard manure (FYM). Inputs should, as far as practically possible, be generated within the farm system.

Including a leguminous crop in the rotation: it is necessary to include a legume to replace the nitrogen used by exploitative crops. Fixing nitrogen biologically from the atmosphere, rather than relying exclusively on bought-in inputs, is therefore used, as far as possible, to satisfy the nitrogen requirements for the rotation. In general, the time that the soil is left uncovered in any particular rotation should also be minimized as far as is practically possible by, for example, including short-term green manures where possible.

Maintaining or increasing soil organic matter: this is more likely to be achieved by observing the first two practices.

Including crops with different characteristics in the rotation: crops that exploit different rooting zones (such as alternating deep-rooting crops

Vetch is an excellent short-term green manure

with shallow-rooting crops) ensure better nutrient circulation over the rotation, help improve soil structure and can ensure that each crop has sufficient area to supply its nutritional needs.

Different crops will also have different competitiveness and weed-suppressing abilities. For example, starting the rotation with a good cleaning-crop, such as potatoes, before moving towards crops that are less competitive against your particular weed populations is usually a good idea.

Other factors to be considered are timing of cultivations, the nature of cultivations, harvest period and intervals between crops. As with so many things in life, the secret to success seems to be variety in all things, but these diverse practices will ensure that the soil profile is continually aerated, nutrients circulated and the weed flora prevented from adapting to any particular regime.

Separating plants from the same families: this is particularly important where crops share pest and diseases problems within the rotation. Alliums, brassicas and potatoes must, in any case, not be returned to the same land within a four-year period (as counted from planting date to planting date). In practice, successional crops (such as spring onions or lettuce) are usually allowed within the same cropping season or twelve-month period. Certain soil-borne pests and diseases can remain active for very extended lengths of time, at times over and above the temporal separations required by standards. The recommended cropping interval for onions in soil with white rot, for example, is at least twenty years.

Exceptional circumstances: in some cases, some of these practices will be difficult to observe practically. For example, many stockless vegetable systems will be in the position that they will need to import nutrients as FYM or green-waste compost, in which case they will normally have to demonstrate to a certification body that they are making maximum use of green manures and legumes within their systems. This can imply that as much as 25–40 per cent of their land can be in fertility building at any one time. Separate rules might also apply to glasshouses or other permanent structures and wild areas and perennial crops are also obvious exceptions to any rules on rotations.

Planning a Rotation

Planning a crop rotation or sequence of crops is the heart of any organic management plan. The various principles in planning a rotation have already been discussed, but the following is a list of the major practical steps to take when planning a rotation.

Define the market: this is perhaps the most important consideration and will require careful calculations to match the area available with anticipated demand. It is dealt with more fully in Chapter 10.

Collect information about the production site: collect as much information as you can about your production areas. This can involve keeping good farm records, as well as gathering information from outside the farm (*see* Chapter 13). This information is also useful in modifying a rotation when necessary.

Choose crops and varieties: depending on the nature of the market and the production site, it is then necessary to choose appropriate crops and/or varieties to grow that match the situation. It is often useful to get advice when planning rotations, especially when converting from a conventional farm system (*see* Chapter 13). Apart from personal advice from advisors or other growers, there is also now computer software – such as 'Orgplan', developed by the University of Aberystwyth – that can help with the design of rotations. However, computer software packages are often not very user friendly and are difficult to interpret, in which case they might need input from an advisor to make the most of them.

Choose fertility-building methods: apart from the actual cash crops, it is also necessary to choose appropriate fertility-building methods. Any fertility-building crops must obviously suit the site and farm situation (*see also* Chapter 4).

Clover contributes nitrogen to a nutrient budget and, as part of a ley, helps improve soil quality

Consider field management methods and practicalities: it is necessary at some point to consider the actual practicalities of implementing a rotation. Experience will also be the final arbitrator of any rotational plant. Whilst there is no definitive guide to practicality, in any situation a whole range of factors can be important. It can be important to vary cultivation methods across the rotation, in order to minimize adverse effects on the soil (*see* below). In some cases, labour supply may vary over the season or be in limited supply in any particular area. It is worth doing a separate chart of peaks and troughs in labour supply-and-demand once the crop rotation is complete, in order to ensure that there is likely to be available the necessary manpower to harvest all those carrots you have careful weeded – or indeed to weed the carrots you have carefully drilled.

Among other numerous important considerations are: variation of sowing times over the course of the rotation (like alternating spring with summer sowing) in order to allow different intervals to tackle different weed populations; synchronizing harvesting with the sowing interval for a follow-on crop or green manure; and the scope for rotating the vegetable area within a whole-farm rotation – for example after one course of the rotation, moving to new field within the grassland on a livestock holding or alternating with arable crops. Above all it is important to allow for flexibility within the rotation. It may, for instance, be necessary to take land out of use to solve certain pest or disease problems, allow time for a bastard fallow if perennial weeds build up, reduce crop densities while fertility builds up or change cropping plans depending on market conditions.

Calculate a nutrient budget: it is useful to calculate a nutrient budget for the rotation once it is completed (*see* Chapter 4 for details) to confirm that it is sustainable. A nutrient budget can help ensure adequate fertility-building periods are built into the rotation, either long-term (one to two year) leys or short-term (as little as six to eight weeks) green manures. It can also help to synchronize supply and demand with individual crops over the whole course of your rotation, as the nutrient sources used in organic agriculture are less immediately available to the succeeding crops, implying that the rotation must try to match release (for instance, from a turned-in green manure) with the crop's peak uptake. Nutrient budgets can also highlight any risk of leaching of nutrients (say, when the soil is left uncovered for more than short periods of time). They may also be necessary to justify the use of supplementary fertilizing, such as FYM, compost and lime. Any nutrient budget drawn up for the rotation can be constantly updated with actual yield figures, in order to compare the actual outcome with the planned outcome, so that any deviations can be taken into account.

Calculate a financial budget: it is also useful to calculate an expected financial budget to evaluate whether the rotation is likely to be financially viable or not. This is discussed in more detail in Chapter 11 where the various ways of doing this are detailed.

Example of a Good Rotation

It is normal for a rotation to initially consist of two to three years of fertility building, followed by three to four years of cropping, although recent research indicates that as much benefit is gained from more frequent short-term green-manure crops as from a long-term fertility-building ley (*see* Chapter 4). It must be added, however, that a longer fertility-building period allows for longer gaps between crops of the same family, which should reduce risks of pest and disease build-up. It also provides a period in which machinery is used with much less frequency and that, in turn, allows longer for biological activity (such as worms and other soil micro-fauna) to improve soil structure. To be balanced against this is the fact that up to a third of the cropping area in the rotation may be out of direct production at any one time, which will affect farm income, especially in the short term.

Apart from effects on soil and fertility, cropping pattern and rotation will obviously also depend on the various farm enterprises being practised. In the case of vegetables, vegetable families are usually rotated in the order: potatoes, brassicas, alliums and roots. Bearing in mind all the factors mentioned above, one example of a rotation that tries to incorporate all the advantages and minimize the disadvantages is outlined in the table below. This also serves to illustrate the many complicating factors involved in implementing a rotation.

Table 7.1 A typical vegetable rotation

Year	Crop	Notes and Observations
1 & 2	fertility building	Grass/clover – for nitrogen fixation. Possible addition of green-waste compost or animal manure if necessary.
3	potatoes	Hungry crop and good for weed strikes and weed suppression once the canopy has developed.
	rye green manure	Over-winter to ensure no nutrients lost to leaching. This is dependent on lifting potatoes in time to sow rye (say by mid-October in most areas).
	add lime	Required every four years or so and is best added in autumn before a brassica crop so the brassica gets the benefit. Also as far away from the next potato crop as possible to avoid scab problems.
4	brassicas	Another hungry crop needs to feature early in the rotation.
5	alliums	Less hungry but also less competitive, so features later in the rotation.
6	carrots	By this stage, hopefully, the weed burden will have been successfully reduced for this relatively uncompetitive and hard to weed crop.

A total of six years, which is the recommended minimum separation time between crops of brassicas and potatoes in order to avoid the build-up of certain diseases

Managing a Rotation

Day-to-Day Management

Probably the most crucial factor to be considered on a day-to-day basis is the prevailing weather and ground conditions. If the soil is too wet, excessive compaction will be caused, which will prevent root penetration, effectively starving the plant as well as making it more susceptible to lodging. A soil suffering from such compaction will often be damaged to such a depth that very deep ploughing (with the loss of much hard-won soil activity/topsoil) or even artificial drainage will be necessary to bring it back into good heart. The efficiency of weeding operations can also be much reduced in wet conditions. On the other hand, cultivating a very dry soil, if it is low in clay content, can cause problems of erosion, as the soil aggregates lack the water they need to hold together and the resulting dust is easily picked up by the wind. This situation is compounded further when rain does fall, as the surface dust is easily washed away from the compacted soil below.

Every activity within a rotation, from ground preparation to weeding and harvest, has its advantages and its drawbacks. These must be evaluated with reference to the relevant factors, including the conditions, the task in hand, the available time and labour, the succeeding crop and cultivation programme. The type of machinery used affects the working depth and the method of soil disturbance. Ploughing inverts the soil, burying the most biologically active few centimetres of top soil at depths where those essential activities become limited, as well as creating an impenetrable pan if a constant depth is used year on year. Rotovating is a vigorous chopping action that destroys much of the larger soil fauna, while harrowing, whilst being less destructive of soil structure, is also less destructive of weeds, creating a further need to drive on the soil during weeding operations. However, a useful phrase to remember is 'shallow turning and deep loosening'. This implies keeping the top few centimetres of soil (where the bulk of the biological activity takes place) as close to the surface as possible in the short term and every few years indulging in a little 'subsoiling' to loosen up the deep layers without turning subsoil on to the surface. Cultivation machinery manufacturers constantly review and redesign their equipment and many new designs are coming on to the market that now follow these principles, often using combinations of discs, shares and tines to keep the soil profile intact.

Careful consideration must be given to the timings of harvesting operations to ensure that the soil need only be driven on in the best possible conditions. This is mainly a question of getting to know the site-specific conditions and learning to adapt to them. For example, wet conditions may render autumn- or winter-sown crops unfeasible or very dry early summers may preclude the late sowing or planting of others. Inevitably these conditions will vary from year to year, so there is little point in rigidly programming a set date to perform any operation. On the other hand, it is necessary to be prepared to go when the opportunity appears. Once again, keeping records of performance in each season will, over time, make it easier to anticipate and forecast some operations fairly accurately.

Monitoring a Rotation

A rotation should be constantly monitored as to its performance and practicality. This implies using the checklist used for planning the rotation as a starting point and looking for symptoms of problems as the farm system develops. Keeping good observational notes and financial records are crucial to this process.

Customizing a Rotation

From time to time it may be necessary to alter a rotation, because for some reason it begins to underperform. Reasons for this can be varied, from inappropriate cultivation techniques, to loss of fertility, to changes in market conditions. It cannot be stressed enough that the two key principles to bear in mind are, firstly, the need to adapt the general rules to your own farming situation (such as marketing opportunities, soil conditions, machinery available, weed, pest and disease problems) and, secondly, the need to innovate and be flexible.

The two most obvious practical considerations are fitting the cropping area to the yield requirements (which may mean subdividing plots for the less popular crops where lower total yields are needed) and fitting crops to the machinery available. There is no point planting 2ha of potatoes if you do not have either the opportunity to sell them or access to the kit to lift them. Fitting crops to machinery also includes adopting standard spacings or row widths that are appropriate to the machinery available. Constantly adjusting machinery settings for different crops is very time-consuming. Sowing dates and deadlines also differ across different areas of the country, varying by as much as four weeks from the mildest areas of the south-west to the colder climes of the north-east.

With time, crops can become susceptible to particular pest or disease pressures and become increasingly difficult to grow successfully. In which case, it may be necessary to start growing new crops, or at least adopt more resistant varieties of the same crop. Again, these pressures will change over time, so that a crop may only have to be avoided for a number of years, after which it can be reintroduced. If localized soil-borne disease is a problem this might necessitate the elimination of a particular crop from certain fields but not from others. Knowing the nature of the fields on a farm is vital in this respect and it is always useful to keep records that include past cropping, any nematode problems and levels, observations on weed burdens, microclimatic and soil characteristics. This is especially important when taking on a new field or farm.

There is often scope to diversify the rotation itself and to diversify within the rotation. Certain marketing situations, notably box schemes and farm shops, often find that they require large quantities of the staple crops, such as potatoes, carrots, onions, leeks and some brassicas, whilst only requiring small quantities of the more labour-intensive crops, such as courgettes and squashes, sweetcorn, salad plants, legumes and herbs. In this situation it is worth considering establishing two separate rotations in two separate areas, one intensive and small-scale and the other extensive and large-scale. Over time it may be possible to rotate the small-scale plot within the large-scale plot, that is to say that once in the course of the whole rotation it moves onto a different plot or plots of the large-scale rotation. Similarly, it is possible to have separate field-specific rotations, if different soil types or other topographical or climatic variations in the different areas indicate that this would be desirable, for instance keeping early crops out of a frost pocket or root crops off a shallow, stony soil.

Within the rotation, diversification on individual plots with the use of intercropping or mixed cropping is possible. Intercropping can potentially maximize yields where space is short, or be used to combine less popular crops in a single course of the rotation. One standard example is the mixing of sweetcorn, squashes and runner beans together; the beans may provide some nitrogen, the corn provides a climbing frame for the beans and the squashes provide ground cover to suppress weeds. The combined yield can surpass the yield of any of the crops on their own. Alternatively, vegetable crops can be intercropped with green manures; a common practice being to under-sow various crops (such as brassicas, sweetcorn, cucurbits) with trefoil or crimson clover, which are relatively low-growing and non-competitive but provide weed suppression and, again, may contribute some nitrogen to the crop. In this case they are left *in situ* after the crop has been harvested, to provide ground cover, and are turned in the following spring, to provide organic matter and nitrogen to the succeeding crop.

Finally, it is crucial to match the cropping plan to the fertility of the soil on the farm and to ensure that the maximum numbers of opportunities for utilizing green manures are built into the rotation. However, in practice, it is unlikely that more than four years of cropping can be sustained from a single fertility-building crop without reliance on bought-in manures.

Troubleshooting

If crops are not thriving in the absence of disease or pest problems (*see* Chapters 3, 6 and 12) – assuming adequate water availability – then it is likely that the problem is either soil compaction or soil-fertility levels.

Compaction

Compaction problems may occur on heavy soil despite precautionary measures, either due to force of circumstances (usually bad weather conditions) or by gradually building up over a period of time.

Precautionary measures to avoid compaction include reducing the need to drive on the soil and encouraging the soil life or biological activity. Earthworms are particularly useful in helping to maintain soil structure. Minimum tillage is also useful in this respect, but is often difficult to implement in temperate organic farming systems. Other techniques can include mulching for weed suppression, raising levels of organic matter, using permanent bed systems, growing perennial crops and interplanting perennial and annual crops. These techniques are of particular interest on heavy soils. Once a compacted layer has formed, subsoiling is the best method of breaking it up. A subsoiler can do an excellent job in the right conditions.

If the problem recurs due to poor drainage then there is the option of installing mole drains (semi-permanent drainage channels tunnelled through the soil by specialist machinery) or permanent drainage. The latter is a very expensive option and rarely lasts as long as hoped for. If the problem is essentially that the land is very wet and/or prone to flooding, then the best option is to use it for other purposes, as the only real long-term solution is to improve the structure and increase the organic-matter content to such a degree that the soil no longer waterlogs. This is achievable in theory, but requires years and years of careful management and dedication.

Deep-rooting crops can also be used to break up compacted layers, as long as they are able to establish well. These can include green manures (chicory in particular), either as a short-term single species break crop or as part of a longer-term ley.

Fertility

If fertility appears to be the problem then it is a question of working more fertility-building measures into the rotation. This can be done by greater use of green manures and nitrogen-fixing crops, both as long-term leys and as overwintering soil protection. Importing sources of organic matter and fertility in the form of farmyard manures, green-waste composts and other agricultural by-products may also be necessary. These can be incorporated or left as a surface mulch, although the former, if done shallowly, will be broken down more quickly. There is a balance to be found between the destruction of soil fauna caused by incorporation, set against the need to release nutrients as rapidly as possible. If your soil is in an impoverished state (as is often the case when starting the conversion process) it may be expedient to hold off from attempting to grow the more nutrient-demanding crops until several years of good management have brought the soil into better heart.

GENERAL NOTE ON MACHINERY

Items of machinery will generally be the most costly individual purchases that a farm or vegetable production business will make. Actual monetary outlay may also be relatively small in comparison with actual running costs for skilled operators, fuel and maintenance, but they are still a significant purchase for which sufficient finance must be found.

Tractors

The most important single purchase will be a tractor, as this will determine to some extent the other types of equipment that can be operated. A tractor may be two- or four-wheeled, depending on the size of holding, and two-wheel drive or four-wheel drive, depending on conditions and the tasks it will need to perform (note that two-wheel-drive tractors are usually cheaper, as they are less in demand). Other factors to consider are wheel size/tracking, hydraulics (front as well as rear, correct type of connection and number of valves for your kit) and whether you want a front-end loader, power steering and cab safety and comfort.

Kit

Apart from a tractor, a wide range of equipment or kit is available for use in organic farming systems for operations ranging from soil preparation, through weeding, to harvesting. The range of horticultural machinery now available is far too extensive to be covered within the remit of this book, so you will need to refer to further sources of information. However, in the last decade or so, some kit, especially for weeding tasks, has been specifically designed for organic systems and new kit is being developed continually. As a rule, it is unlikely that a farm will be able to afford all the kit that they would like and priority should be given to equipment that will be most used. Factors can include:

- generalized versus task-specific: which piece of kit is going to be the all-round most useful for your cropping plan? To what extent can that

A tractor, although expensive, can perform a variety of tasks for many years

piece of kit be adapted to perform tasks other than the one it was necessarily designed for (for example, rotovators can be used for primary cultivation and muck spreading, as well as for the secondary cultivation for which they were designed)?

- suitability: to soil type/conditions/holding size
- adaptability: of the new machinery to the spacing/row-widths/bed sizes already in use
- repair and maintenance: how much time and know-how do you have available? How long can you afford for machinery to be out of use? Do you know a reliable local mechanic?
- budget available: cost is obviously the prime consideration, though you may be able to get a good deal for new kit if you buy more than one item.

Obtaining Machinery

Second-Hand vs New

It may be possible to purchase second-hand kit at a cheaper price. Most areas have local auction houses that, over time, offer a wide range of machinery. Most have a catalogue mailing-list that can be subscribed to and the availability of specific pieces of kit is likely to be a matter of chance. It is necessary to be aware of the potential weak points in any piece of equipment so as to be able to spot problems and it stands to reason that everyone else there is also looking for a bargain and competition may be fierce. The fact that auctions abound with dealers cuts both ways; they are serious competition but are also an excellent source of second-hand machinery and it might be more cost-effective to allow the dealer to do the searching.

Machinery Rings

It may be possible to borrow or lease the necessary kit if purchase is not an option. Some areas have machinery groups or 'machinery rings' that own or pool items of kit (or even labour in some cases). Each member can book the time they need or pay for it on a commercial basis, depending on the system in operation. The disadvantage of these groups is that it can be difficult to get access to machinery at exactly the time that it is needed on-farm.

Contractors

Before making any major purchases it is worth finding out how much local contractors charge for less frequently performed jobs, such as muck/limestone spreading or hedge-trimming, as they may be more cost effective than purchasing your own kit. Similarly, it may be possible to hire or borrow from neighbours or to purchase collectively. If a machinery group does not already exist in your area it is worth trying to set one up.

Other Considerations

Also bear in mind that if you move from conventional to organic land with the same piece of kit, it will be necessary to clean it down between uses to ensure that no conventional fertilizers, pesticides or their residues in the soil are carried on to the organic land. If your budget stretches to it, it is worth considering purchasing dedicated machinery for each type of land/crop.

Chapter 8

Protected Cropping

Protected cropping is an important part of UK organic production, accounting for 26 per cent of the wholesale value of all UK-produced organic vegetables in 2003–04. The facility to extend the season at either end is becoming vital for organic growers. It can improve cash flow, maintain continuity, command higher prices for crops, increase the range of crops that can be grown and can help to minimize the quantities of bought-in or imported produce. It also allows the growing of more exotic, higher value, niche crops. All these can be especially important for farmers and growers selling directly to the public through box schemes, farmers' markets and farm shops. In addition, the workload on a holding can be spread over a longer period and work can be found inside when inclement weather does not permit activity in the field.

Apart from growing crops, there are many other potential benefits to having greenhouses or polytunnels on a holding. For instance, they can be useful for plant-raising, drying crops such as onions and garlic, chitting potatoes and, on mixed holdings, for lambing sheep and raising ducks and turkeys. While important for UK production, spreading the season and reducing food miles, they are not without their critics and the rapid growth in the use of Spanish tunnels in areas such as Herefordshire has caused concern to local residents over their visual impact on the countryside.

TYPES OF PROTECTIVE CROPPING

Crop Covers

The use of crop covers, such as fleece, have become popular with organic growers as a cost-effective way of extending the season by protecting sensitive crops from frost and raising the temperature of the growing environment around the crop. They also warm the soil, thereby encouraging mineralization of nitrogen. In all, they can advance crop maturity by 10–14 days.

Fleece is a lightweight non-woven porous material that is used as a floating mulch. It is supported directly by the crop and has the advantage that it can also provide barrier protection against insect pests, such as aphids, carrot fly and cabbage root-fly, as well as larger pests such as birds (*see* Chapter 6). The disadvantage of fleece is that it is not transparent to the eye and monitoring the growing crop underneath is not easy. Weed growth is also encouraged under fleece and the task of removing the covers to allow physical weed-control is labour intensive. Fleece is probably better at extending the season in the spring rather than in the autumn as, although it can protect crops against up to 5°C of frost, it encourages soft growth, which can reduce winter hardiness of covered crops.

Wide rolls of 10–12m are normally used, but it can also be bought in bed widths. Fleece comes in different weights, from 17–30g/sq m. Although fairly cheap, it can be damaged relatively easily by deer and other animals, particularly the lighter-weight covers. Realistically and practically, it can usually only be used for one or two seasons, especially if the pest-exclusion aspect is important. If stored for re-use it should be kept out of the way of mice, as they love to build nests in the rolls!

Other crop covers, such as perforated-plastic mulches, which are used in conventional carrot-production, are less appropriate for organic systems because of lack of airflow and problems with weed growth. Mesh-netting covers are useful for pest control, but they have little effect on crop maturity or frost protection.

Cloches/Low-Tunnels

The use of fleece and Spanish tunnels has meant that cloches and low tunnels have become virtually redundant. The use of hoops over beds can be useful where crops have a tender, exposed growing-point, such as peppers, and would be damaged by wind abrasion from fleece. They offer similar benefits to fleece, but are more labour-intensive to work with.

Polytunnels

Polytunnels are metal-hooped structures covered in plastic, which provide a cheap and affordable alternative to glass (*see* below). Plastic covers behave differently to glass, as the air inside a polytunnel both heats up and cools down more quickly than inside a greenhouse. Overall temperatures reached are not as high and on clear, frosty nights temperatures inside can be as low as those outside.

Polytunnels come in many shapes and sizes, but the most familiar and widely used by organic growers are the traditional single-span and multi-span tunnels. Single-span tunnels usually range in width from 2.44m (8ft) up to 9.14m (30ft) with variable lengths. Generally speaking, wider tunnels are better. However, excessive lengths should be avoided

Polytunnels are a useful addition; extending the season and expanding farm enterprises

– over 30m (100ft) – as hotspots can form in the middle of the tunnel (unless the sides are vented). Straight sides are also an advantage, as they enable a greater proportion of the growing area to be used and do not restrict the growing of low crops to the outside beds. Many tunnels are designed to withstand strong winds and tunnels with wider-diameter tubes and extra bracing are recommended for more exposed sites.

It is worthwhile taking care to choose the best location for tunnels. Either choose a site sheltered from winds or erect windbreaks. However, excessive shading and shelter that allow no air movement through the tunnels should be avoided. Alignment from north to south generally works best, to avoid shading effects of taller crops. A slight slope from end to end works well. In a mixed holding it is often most convenient to locate the tunnels near to the farmhouse for ease of management, access to water and electricity and for security. Other considerations may include avoidance of frost pockets, prior fertility building, good soil structure and absence of perennial weeds. Before erecting tunnels it is best to get professional advice, but key points include making sure that the foundations and tunnel structures are secure, ensuring that there is adequate provision for doors and that the plastic is firmly stretched over the tunnel hoops. The tighter the fit, the longer the cover will last.

For smaller growers, single-span tunnels have the advantage that different crop-groups can be isolated from each other and managed separately, the tunnel environment being manipulated according to crop requirements. They are also cheaper to purchase and can often be picked up second hand. Multi-span tunnels are, as the name suggests, two, three or more single polytunnels joined together into one large structure. These are more expensive to purchase and usually require erection by specialist companies. They can include rain gutters that are used to tension the polythene covers. The larger size of these structures makes it easier to use machinery inside.

Ventilation is the key to healthy crops and designs of both single- and multi-span tunnels are becoming evermore sophisticated for this purpose, with roll-up roofs and fully ventable sides and ends. Side ventilation is common in multi-spans, using a roll-up blind for the bottom metre or so on each side. Fine-meshed plastic nets on this lower section can reduce wind speeds over the crop and may prevent scorch due to excessive transpiration. Recently, ventilation systems have become available for multi-span tunnels, though they do not come cheap and may only be necessary in the centre of the largest structures and, even then, not in every bay.

Spanish Tunnels

The low-cost, moveable, multi-bay Spanish tunnel has seen rapid take-up in the past five years or so, mostly as a means of extending the strawberry-growing season. They are also suited to high-value vegetable crops. Spanish tunnels are similar to polytunnels, but the tunnel structure is usually of much lighter construction and not permanently fixed with foundations. They can also be opened at the sides. As compared to polytunnels, the airflow is improved, which reduces humidity and potential disease problems, and the large air-volume means a more stable climate. They can be fully opened to ensure crops do not overheat in the summer or to allow exposure to frost in the winter. With heights of between 3.5m (11.5ft) and 4.5m (15ft), access for field tractors is easy.

The versatility of Spanish tunnels often appeals to growers, as they are temporary structures, generally not needing planning permission, and can be erected at the start of the season and removed at the end. They can, therefore, be rotated around a farm with less need for a rotation within the tunnel itself. There have been improvements in design since they were first introduced, to improve strength and resistance to winds and reduce labour needed for skinning. Guttering can remove problems of waterlogging in the valleys between the tunnel bays. The ability to interlink tunnels provides further structural stability and there is no wasted ground between tunnels. Telescopic tunnels are now on the market, with extendable legs that enable the tunnels to be lowered early in the season – when growers need to extend the season – and lifted later in the season to allow larger air-volumes and better ventilation.

Location requirements are similar to traditional tunnels, but they should be positioned lengthways to the prevailing wind, as they are capable of withstanding stronger winds when it blows lengthways down the tunnels rather than sideways across them. A 50 per cent permeable windbreak can offer better protection than a dense windbreak. Good drainage is also important, avoiding wet areas, as rainwater is concentrated into the leg rows. Tunnel manufacturers may provide training in construction techniques and a technical back-up service.

The biggest management task is venting. Spanish tunnels can be completely opened to ensure the crop is not stressed by heat or humidity. Computer systems with alarms and text-message warnings are now available, which will alert tunnel managers to different critical thresholds of wind, temperature and humidity, that can be set according to the requirements of the crop. The vents can be used to manage high winds, although at wind speeds of over 110km/hr (70mph) the polythene should be removed and tied into the leg row.

Greenhouse Cropping

Organic production in glasshouses comprises a very small sector of the industry. Glasshouses have an advantage over polytunnels in that they are more efficient to heat, as heat retention is better. However, the capital

A glasshouse can be useful for plant raising and to extend the season

set-up and maintenance costs are high. A small- or medium-size greenhouse can be useful for plant raising and extending the season further than is possible with polytunnels. Greenhouse cropping can permit the growing of longer-season crops and extend the season for crops, such as tomatoes, peppers and cucumbers, to periods when their price should be high.

High investment costs and the length of the conversion period (up to two years) remain an obstacle for conversion to organic glasshouse production, as it is difficult to justify the loss of income during the conversion period. In-conversion cropping is not recommended by certifying bodies after a period of intensive cropping, yet fertility building in a glasshouse situation is an expensive option. A better option might be to erect new glass onto land that has already been converted and has had a period of fertility-building.

Solar Greenhouses

These are popular, mainly with small growers, in the United States and are super-insulated houses designed to collect and retain solar energy. The simplest example of this is a lean-to greenhouse on a large-garden or smallholding scale. Solar greenhouses differ from conventional greenhouses by having glazing oriented for maximum solar heat-gain during winter. The long axis runs from east to west and the south-facing wall is glazed to collect the optimum amount of solar energy, while the north-facing wall is well insulated to prevent heat loss and covered or painted with reflective material. They have a steeply pitched north roof. Active solar greenhouses use supplemental energy to move solar-heated air from storage or collection areas to other regions of the greenhouse, usually through pipes in the floor. This could be suitable for a propagation house.

Plastic Greenhouses

An alternative to glass, and intermediate between a greenhouse and a multi-span tunnel, is a greenhouse covered with bubble plastic. Less expensive than traditional greenhouses, they are claimed to have excellent heat-insulation and light-diffusion properties, eliminating shade and hot spots, and to give up to 50 per cent fuel saving over conventional un-insulated greenhouses.

CROPPING

Rotations

The rotation requirements of the organic standards are less rigid for protected cropping systems than for arable and horticultural crop rotations, but the principles, as outlined in Chapter 7, are the same. Crops must

Many different crops can be grown in polytunnels

be grown in the soil to qualify for organic status and mono-cropping or annual cropping of the same genus is permitted (with the exception of alliums, potatoes and brassicas). This allows the continuous growing of crops such as tomatoes. For the smaller-scale grower wishing to increase the variety and value of crops grown, a fuller rotation is possible and desirable, although many of the popular crops divide into the Solanaceae family (tomatoes, peppers, aubergines) or Cucurbitaceae (cucumbers, melons, early courgettes).

It is difficult, economically, to include longer-term periods of fertility building within a polytunnel or glasshouse rotation, but shorter-term green manures can be included and taller crops, such as tomatoes, cucumbers or climbing French beans, can be under-sown with a low-growing white clover or trefoil. Bear in mind that these cover crops will also need monitoring and management, for example mowing and irrigation.

The other requirements of polytunnel space also need to be taken into account when designing a rotation. Drying onions and garlic can be an important use for tunnels in late August or September. Pumpkins and squashes can be cured inside if the weather does not permit it in the field, though they are not suitable for longer-term storage due to the risk of frost damage. If raising bare-root transplants, as opposed to modules, then this can also be fitted in to the rotation, though for brassicas this could be a problem if winter-salad production is planned. Good ventilation is also necessary, to reduce downy mildew risk on brassica transplants.

Crop Choice

There are three main periods (with some overlap) for protected cash crops: spring, summer and autumn/winter. The spring to early summer period can be a profitable one, as produce from the field crops can be in short supply (the hungry gap) and customers are seeking new and exciting tastes after winter staples. Early carrots, beetroot, radishes and turnips for bunching can also be good crops. Other options for this period are sugar snap or mangetout peas, lettuce and other salads. Climbing French beans, early celery and courgettes traverse the spring period into early summer and can be cropped until the outdoor harvest is in full swing. Early leeks could be grown in the spring prior to a summer crop. Table 8.1 provides an example of the range of crops that can be grown in a typical protected cropping system supplying enough for fifty customers per week.

During the late spring/early summer, once the winter and spring crops are cleared, the tender frost-sensitive crops can be planted. These include tomatoes, peppers, aubergines, cucumbers, melons and basil. Some growers feel that these summer crops, with the possible exception of basil, are not the most lucrative. For example, it is easy to over-produce tomatoes and cucumbers, which tend to crop most heavily under un-heated plastic at the same time as many customers are taking their annual holidays or are harvesting their own. The prices, therefore, tend to slump at this time. It is possible to process some of this produce, such as tomatoes into chutneys or sauces, to add value and to help to use up a glut.

In recent years the growth of popularity of mixed-salad bags has provided a boon for organic growers, as many of the constituents can be grown easily and profitably over the winter period. Until recently the only viable option for growers was to grow lettuces over this period, with all their attendant problems, such as downy mildew and fungal rots. However, it is currently possible to source a wide range of other leafy salad vegetables, which can be drilled (or transplanted in modules) in the autumn after the summer crops are removed in October or early November. They can then be harvested from November onwards through to the spring, although there is often a gap of a few weeks during cold weather in January. Leaf growth increases rapidly in February once day length and temperatures rise, until the crops go to seed. The main issue is that many of these salad crops, such as the oriental greens (mizuna, green-in-snow, mibuna, komatsuna, Pak Choi, Chinese cabbage, red mustard), rocket and land cress are brassicas and, thus, subject to rotational

Table 8.1 Example of a weekly produce requirement from polytunnel cropping for a box scheme or market stall serving fifty customers

Crop	Unit size	Quantity (50 cust)	Frequency	Sales period	Average weekly req.	No. of weeks	Total req. for season	Yield per plant	No. of plants	Plants per sq m	Total area in sq m
Tomatoes	0.5kg	25kg	weekly	Jun–Oct	25kg	20	500kg	2.5kg	200	2	100
Aubergine	400g	20kg	monthly	Jun–Oct	5kg	18	90kg	1kg	90	3	30
Cucumbers	1	50	weekly	Jun–Oct	50	20	1,000	20	50	1	50
Peppers	300g	15kg	fortnightly	Jun–Oct	7.5kg	18	135kg	1.5kg	90	4	22.5
Cl. French beans	400g	20kg	weekly	May–Jul	20kg	8	160kg	2.5kg	64	10	6.4
Onions	500g	25kg	weekly	Jul–Sep	25kg	12	300kg	250g	1,200	30	40
Courgettes	500g	25kg	fortnightly	Jun–Jul	12.5kg	4	50kg	2.5kg	20	1	20
Parsley	50g	2.5kg	weekly	Jan–May	2.5kg	16	40kg	400g	100	20	5
Basil	60g	3kg	fortnightly	Jun–Oct	1.5kg	18	27kg	270g	100	20	5
Salad leaves (1)	100g	5kg	weekly	Sep–Mar	5kg	26	130kg	500g	260	10	26
Salad leaves (2)	100g	5kg	weekly	Sep–Mar	5kg	26	130kg	500g	260	10	26
Salad onions	bunch	50 bunches	fortnightly	Feb–Jun	25 bunches	16	400 bunches	0.2 bunches	2,000	400	5
Spinach	200g	10kg	fortnightly	Apr–Jun	5kg	10	50kg	500g	100	16	6.25
Early beetroot	bunch	50 bunches	fortnightly	May–Jun	25 bunches	8	200 bunches	70g	1,000	80	12.5
Early carrots	bunch	50 bunches	weekly	Apr–Jun	50 bunches	10	500 bunches	50g	4,000	250	31.25

(1) Brassicas: rocket, mizuna, tatsoi, mibuna, red mustard, green mustard
(2) Non-brassicas: Claytonia, chrysanthemum greens, Chinese celery, cornsalad
Table courtesy of the Organic Advisory Service

Table 8.2 An example of a work schedule for polytunnel cropping

Crop	**Jan**	**Feb**	**Mar**	**Apr**	**May**	**Jun**	**Jul**	**Aug**	**Sep**	**Oct**	**Nov**	**Dec**
Tomatoes	S	S			P		H	H	H	H		
Aubergine	S	S			P		H	H	H	H		
Cucumbers			S		P	H	H	H	H	H		
Peppers	S	S			P		H	H	H	H		
C.F.Beans			S	P	H	H	H					
Onions						H			P			
Courgettes			S	P		H						
Parsley	H	H	H			S		P		H	H	H
Basil			S		P		H	H	H			
Salad						S1	S1	P1	S2		P2	P2
Leaves	H	H	H						H	H	H	H
S. onions	H	H	H	H	H			S	S			
Lettuce			H	H			S	S	P	P		
Spinach		S	S	H	H	H						
Beetroot	S	S			H	H						
Carrots	S	S		H	H	H						

S = sowing time, P = planting time, H = harvesting time
Adapted courtesy of the OAS

restrictions. Other non-brassica salads that can be grown for salad bags include lamb's lettuce or corn salad, winter purslane or claytonia, endive, chicory, perilla, baby-leaf spinach and leafy varieties of lettuce. Other over-winter crops that are possible include parsley, spring onions, garlic (for fresh sales) and over-winter onions.

When designing a rotation for a polytunnel it will be necessary to plant to spread the workload over the cropping calendar and crops should be chosen with this in mind. Table 8.2 shows a work schedule with a range of crops that effectively spreads the labour requirements over the season.

Soil Fertility and Management

Organic standards state that crops must be grown in the soil and, thus, hydroponic and grow-bag systems are not permitted (apart from herbs and ornamentals). This is a major contrast with much conventional protected-cropping. Manures and composts are likely to be the principal sources of soil fertility. Manures should preferably be sourced from within the farm but if not, and if from a conventional source, will need to be composted for three months or stacked for six. A good reason for using composted manures is that the requirement for an interval between application and harvest is reduced from three months to two. This period can be further reduced if proper composting can be demonstrated, with analysis of the end-product compost to ensure that no human pathogens are present. An increasing body of evidence indicates that the use of compost can also have a suppressing affect on plant diseases. Properly composted manures or green-waste compost should also be free of weed-seed contamination. If space allows, it is useful to have an area for composting reasonably close to the tunnels, as they can produce plenty of organic material for composting, such as tomato haulms. In contrast with field production, there is no upper limit (of 250kg N/ha) that can be applied to the land area and manures or composts can be applied at any time of the year.

Supplementary feeding may be needed for nutrient-demanding crops, such as tomatoes. Only those liquid feeds derived from permitted organic and mineral materials may be used and not those based on soluble salts. Fishmeals and fish emulsion (free from non-permitted substances) are allowed in protected cropping but not in the field. These and other feeds made from seaweed extracts or concentrated cow manure come into the restricted category and permission from the certifying body is required prior to use. Liquid feeds can be applied through a diluter into the trickle irrigation. Home-made feeds can be produced by fermenting the leaves of nettles and comfrey, though difficulties of application in the irrigation water make it practical only on the smaller scale. It is important to avoid contact of the feed, particularly home-made feeds, with the plant leaves or fruits that are to be eaten.

The use of a pedestrian or tractor rotovator is probably the commonest form of soil cultivation in tunnels. Permanent raised beds with minimal disturbance are also feasible, after initial deep cultivations. Regular soil-analysis is recommended, to monitor if fertility levels are being maintained and to be sure there is no build-up of mineral-salt levels.

Irrigation

High temperatures and rapid crop growth under protection mean that demands for water can be high. Watering regularly and often is important, so that crops do not become stressed. Over-watering or uneven watering is also a danger that can lead to problems, for example with splitting in tomatoes. The basic choice is between overhead irrigation using sprinkler lines or drip irrigation. Sometimes a combination of the two can be used.

Sprinklers are useful for drilled crops, but overhead irrigation can raise humidity levels and lead to fungal problems on crops. Water marks on foliage or fruit can also reduce crop quality. Drip or trickle is more efficient in the use of water, as the moisture is targeted to the root zone, particularly when used under plastic mulches. It does need monitoring for blockages to the holes and damage to the hoses, say from mice.

Climate Control

Temperature and humidity control in tunnels is primarily by ventilation through opening the end doors, or panels in doors, and side panels, if present. Some crops, such as cucumbers, have higher humidity requirements than others. Although this can be automated, at a cost, regular monitoring and adjustment should suffice. During the spring and autumn it is important to shut the tunnel at night to reduce heat loss. Temperatures can rise rapidly on a spring day inside the tunnels to 40°C or more and crops can easily be overheated if tunnels are left closed by mistake. In glasshouses, more sophisticated computer-controlled ventilation systems are often used.

Pests and Diseases

The high intensity of cropping in warm and humid environments can lead to high pest and disease pressure. In greenhouses, lack of rotations and a tendency towards monocropping means that regular crop monitoring is essential and training in pest and predator identification may be needed.

Pest control in protected cropping is possible using biological controls, as high temperatures and an enclosed environment allow the pest predators and parasites to function well. Biological controls are available for aphids (various parasitic wasps, gall midges and fungal spores), caterpillars (Bt), leafhoppers and leaf miners (parasitic wasps), red spider mite (predatory mites and midges), sciarid flies (nematodes), thrips (predatory mites), vine weevil (nematodes), whitefly (parasitic wasps, ladybirds and fungal spores) and slugs (nematodes). Regular monitoring of crops is essential, as it is important that predators are introduced early before pest populations have built up too high. Natural predators can also be encouraged by companion planting of attractant plants, around the doors of the tunnels or amongst the crops, using phacelia, marigolds, fennel or coriander (some of an early crop could be allowed to flower, to encourage hoverflies). Basil can be grown in the same tunnel as or underneath tomatoes and may reduce pest attacks. Ducks can be used in polytunnels or glasshouses between crops to clean up slugs and snails. Some growers integrate small ponds or water features into their tunnels in order to encourage frogs and toads.

Good ventilation and minimization of overhead irrigation is necessary to avoid high humidity, which can encourage fungal and bacterial diseases.

The choice of polytunnel film can also help, by reducing drips onto plants, and changing the light spectrum may reduce botrytis (*see* below). Many pot-based studies have shown that the use of compost can have a suppressive effect on soil-borne diseases, such as damping-off and root rots (*Pythium ultimum, Rhizoctonia solani, Phytopthora* spp.) and wilts (*Fusarium oxysporum and Verticilium dahliae*). Steam sterilization or pasteurization of soils may be used in protected structures, with permission by the certifying body, as a 'one-off practice' to combat a particular pest problem. It should not, however, be a regular part of the husbandry system. Work has shown that low-temperature/short-duration soil steaming could become a sustainable alternative to high-temperature soil-disinfection.

Glasshouse and polytunnel hygiene is important. Power-washing of glass and the use of biodegradable soaps to clean fixed structures is acceptable. Weeds can act as over-wintering sites for pests in both polytunnels and glasshouses.

Weed Control

Weeds can grow very quickly under protected conditions and careful monitoring and management is essential. As within any vegetable-production system, it is vital to remove any weeds before they set seed.

Mulching with various materials is a useful method of weed control in protected cropping. The easiest method of weed control in tunnels is to use black plastic mulches, in combination with drip irrigation so that the pathways between beds remain dry. Biodegradable plastic mulches of various types are now available. These come in two grades, one with a maximum life expectancy of sixteen weeks, suitable for mineral soils and rapidly growing crops, and one with an expectancy of twenty-six weeks, for slower-growing crops and more microbiologically active soils. Although degradation may be faster under tunnel or glasshouse conditions, there would not be the same problems from wind that can occur in the field.

Mechanical weeding using inter-row equipment may be possible in glasshouses and multi-span tunnels, which allow the use of small tractors. Otherwise it is down to conscientious use of hand hoeing and hand weeding, making sure only to use draw-hoes on the beds nearest to the polythene! These edges can be tricky to manage and can easily become refuges for perennial weeds, such as docks. Heavy mulching of the area between adjacent tunnels can help, in this regard, to prevent invasion of weed roots from this area.

Stale seedbeds are possible for drilled crops, if time permits, preparing the ground for the crop and irrigating to allow a flush of weeds. Avoid the use of flame weeding in plastic polytunnels!

OTHER CONSIDERATIONS

Planning Permission

It is important to check with your local planning authority as to whether planning permission is needed for protected cropping, as requirements can vary between different areas. Normally, planning permission from the local planning authority is required where the structure is 'permanent' on non-agricultural land. In certain circumstances, permanent polytunnels on agricultural land may require the prior notification of the local planning authority under the requirements of the General Permitted Development Order 1995 (GPDO), providing they meet certain criteria that apply to other agricultural buildings. A permanent polytunnel over 465sq m (5,000sq ft) on an agricultural holding over 5ha (12 acres) in size, would usually require the benefit of planning permission. Under this size, prior notification only might be required and, if it is not permanent, may not require permission at all.

It is essential to take professional advice on these matters before investing any money, as 'permanence' appears to be open to interpretation and this can vary between planning offices. In general, the process of fixing a building to the ground creates the permanent status and thus promotes a temporary activity to 'operational development', which often requires the benefit of planning permission.

Tunnel Covers

The range and scope of tunnel covers available to growers has increased rapidly over the last few years. The most basic tunnel covers available are guaranteed for four years and many can last for longer, especially with the use of anti-hotspot tape on the hoops. Many offer special properties, such as improved heat retention at night, anti-drip films, long-life and reduced heat-build-up during the day. Some claim to reduce the disease incidence of botrytis and other diseases. The choices available are:

- Standard polythene (EVA/UVI). This is the cheapest option: a clear, polythene film, suitable mainly for growing frost-hardy crops and/or growing summer vegetable crops, where condensation is not a problem. These covers are expected to last for at least four years.
- Thermal/anti-fog films. These covers give improved heat retention due to an infrared additive in the polythene and are suitable where tunnels are to be heated at night (say, for plant-raising) and for over-winter vegetable production. The anti-fog films have additives to control the condensation of water on the film. Water will condense uniformly, rather than as droplets, reducing risks of drops falling on growing

plants and improving light-transmission. This can help reduce fungal diseases such as Botrytis. It is best for crops that respond well to high direct-light levels and warm temperatures. Covers should last for four or five years.

- High light-transmission films. These covers are designed to make more light available to plants by increasing the amount of diffused light and reducing the proportion of infrared light. They can raise tunnel temperatures by a degree or two in the early spring and also cool the tunnel environment by a similar amount later in the season. Good growth can be achieved without risk of scorching. Thermic properties can help to maintain overnight temperatures.
- Anti-grey mould (Botrytis) films. These covers block the entire ultraviolet (UV) spectrum, allowing enhanced blue-light transmission. It is claimed that these properties inhibit the growth of Botrytis by suppressing spore production.
- In addition, other optional extras that can be included in the films are: increased absorption of UV light to reduce insect infestations and fungal diseases; and transmission of short-wavelength UV for colour, fragrance and flavour development. White tints and diffusion options are available, but mostly used for nursery stock. Research is in progress on the use of anti-glare and coloured films.

Recycling of Plastics

The disposal of horticultural and agricultural plastics is a big issue and burning of plastic waste is prohibited within the organic standards. There are now a number of recycling schemes around the country that operate collection services for a subscription cost and/or collection fee. The Soil Association (SA) has a waste-minimization officer and information is available on their website or from DEFRA.

Chapter 9

Harvest, Transport and Storage

Harvest and post-harvest handling of vegetables can have a large effect on the quality of produce and will obviously affect the proportion of the crop that can be marketed and the price that it will attain (*see* Chapter 10). There are a large number of issues surrounding harvest, storage and transport of vegetables and these include organic standards, as well as technical, environmental and health and safety concerns. This chapter will provide a brief overview of the main points.

All growers need to harvest, handle and transport vegetables, even if only from the field to grading line or pack house. In some cases, growers will need to transport produce some distance to wholesalers, packers or even consumers. During this process, care needs to be taken not to handle produce excessively and increase the risk of damage or deterioration.

Currently, relatively few organic growers store vegetables for long periods of time. Holding produce can allow growers to avoid glut periods and so attain higher prices in the marketplace and help maintain continuity, especially with direct sales. However, offset against this must be the costs associated with storage and extra handling. There might also be a price to pay with respect to freshness and taste of the product. In general, organic crops can be stored using the same methods as for conventional crops. There is no evidence that organic crops store less well than conventional crops under the same conditions, but there might be an increased risk of higher losses because pesticides and sprouting suppressants cannot be used. However, problems can be reduced by storing healthy, undamaged crops under the appropriate conditions and by using good organic husbandry techniques before harvest.

REASONS FOR STORING CROPS

Organic growers give a range of reasons for storing crops. These vary from situation to situation and include:

- providing regular income and avoiding cash-flow problems
- spreading workload throughout the year (to maintain staff)

- achieving continuity of supply to customers, especially where direct marketing is used
- increasing quality of the produce over a longer period
- increasing consumer confidence by ensuring that they are receiving local produce over a longer period (as opposed to bought-in top-ups)
- avoiding oversupply and market saturation at peak harvest time
- accumulating produce for peak periods of demand. For example, short-term cold storage of perishable crops can enable more efficient use of labour, as several days' supply can be harvested at the same time. Also bought-in produce (if used) can be bought in larger batches and kept in good condition until it is sold on.

Where organic growers have been questioned, they also gave various reasons why storage of crops is not necessarily a viable option. Once again, these vary from situation to situation and are likely to change over time, but include:

- lack of incentive – many growers are able to plan for and market all their produce as it is harvested
- economies of scale – small businesses are unable to generate the funds necessary for capital investment in sophisticated harvest-rigs or storage facilities
- a wide range of crops are grown – the variety of crops on a holding and the different market outlets characteristic of many organic farms complicates the provision of adequate storage facilities that satisfy all needs
- lack of technical and financial information – there is little information on the fluctuation of wholesale market prices during the year and this makes it hard for growers to decide whether it will be worth storing produce or not.

It is likely that all growers will be faced with transporting and storing some produce, at least for short periods of time. The rest of this chapter aims to outline the various methods growers can use to meet this challenge.

AVOIDING STORAGE PROBLEMS

General Practice

There are a number of existing organic management practices that can help to prevent transport and storage losses and that are generally an integral part of any normal organic husbandry programme. Storage problems can be avoided completely by planning the crop cycle and harvested quantities to coincide with demand periods. In this case, it is only necessary to harvest, handle and market the fresh produce. However, if planning to store produce, it is important to choose varieties that are known to travel well and have a longer shelf life (*see* Chapters 3 and 12). Disease-free seed should also be used and care taken to prevent pest and disease attack on the crop in the field (*see* Chapter 6), as this can adversely affect produce quality in storage. In addition to this, it is necessary to avoid oversupply of nutrients (especially nitrogen, *see* Chapter 7) and undersupply of others (for example calcium). Erratic irrigation should be avoided to prevent skin cracking on vegetables.

Harvesting

It is important to harvest produce at the correct time and under the appropriate (cool and dry) weather conditions. It is inadvisable to harvest in the rain, which can lead to drying problems and the risk of storage of wet crops. Conversely, in sunny conditions, produce should rapidly be transferred to cooler or shady locations. Produce harvested at the physiologically appropriate stage is likely to store (and transport) better than over-mature vegetables or fruits.

At harvesting, great care should be taken to carefully handle and grade the produce to avoid any damage to the crop at this stage. To this end, methods of minimizing handling should be sought. These might include harvesting, grading and packing vegetables in the field, on mobile rigs

Special harvesting-rigs are available that minimize handling at harvest

designed for the purpose. Grading by hand is usually gentler than grading mechanically (but more time consuming). Health and safety regulations also need consideration at this time. For instance, observe the required period between applying composts or manures and harvesting vegetables and use clean equipment at harvesting.

Crops to be stored should be free from skin damage, bruises, spots and rots. All these can lead to rapid moisture loss (and poor appearance) or provide entry for organisms that cause decay. Vegetables may need to be washed to remove debris or soil. When trimming and cleaning vegetables, at least some material can often be left to protect or cushion, and ultimately present, the marketable part of the plant. When stacked, crates should not be overfilled, allowing produce to fall out or be crushed. If repacking from crates on grading lines or for boxes, care must be taken not to bruise or otherwise unnecessarily damage vegetables.

Handling and Sanitation

Health and safety issues come to the fore when handling and preparing produce for marketing, principally to protect consumers from either poisoning or food-borne illness. Maintenance of integrity of organic produce must also be a prime consideration, so that organic produce can be clearly identified and protected from contamination during any harvesting, storage or transport processes.

Washing can reduce the incidence of post-harvest diseases (and thus maintain appearance and storage life) of fresh produce. If washing vegetables, only water (liquid, steam or high pressure) can be used or organically approved cleaning agents, such as hypochlorite in solution. In this case, produce should be rinsed after washing. Vacuum cleaning is also permitted.

Any storage structures should be clean and monitored to make sure that produce enters and leaves in a timely manner and, if necessary, is rested to allow pests and diseases to die off. Temperatures and humidity should be monitored and airflow maintained.

When storing fresh or harvested produce in enclosed spaces, the two most important factors are temperature and humidity. Basically, the cooler the temperature (within reason) the longer produce can be maintained, as respiration, aging (for instance, ripening or softening), moisture loss, spoilage due to micro-organisms and growth are all slowed at lower temperatures. Freezing is not, however, normally suitable, unless the produce has been pre-prepared in some way. Temperature should, therefore, generally be 1–2°C or above (depending on the vegetable) and maintained at a constant level, to avoid condensation of water on produce (which can lead to deterioration and rots).

Washing can be important for appearance and helps prolong quality during storage

High humidity (RH 85 per cent or more) is important in reducing moisture loss from vegetables and maintaining their appearance. This is often difficult to manage at the same time as maintaining low temperatures (as refrigeration removes moisture) and preventing diseases developing (which prefer higher humidity). Humidifiers exist to maintain humidity in stores where this is important.

Transport and Packaging

Transport to pack houses or other intermediaries should be done in conditions and containers suitable to the product, to avoid unnecessary damage and potential rejection of produce. Transport should be done in sealed containers or packaging that does not allow the contents to be substituted or contaminated. All containers and sacking should be of food-grade quality and free from any residues. Vehicles should also be cleaned regularly to avoid any contamination and always after they are used to transport any conventional produce.

Packaging is important, both for marketing and maintaining the integrity of organic products, but an excess of packaging should be avoided. Ecologically sound materials should be used, as detailed in the organic standards, and non-essential packaging avoided if possible. If distributing vegetable boxes, these should be packed accordingly, with the heavy items at the bottom and the more delicate items at the top. They should also be transported in a sensitive manner and not left standing for long periods in hot vehicles (especially in the summer).

STORAGE METHODS FOR ORGANIC VEGETABLES

Storage methods vary depending on the produce to be stored, the intended market outlet, the length of time of storage, the farm facilities and, ultimately, the quantities and value of the crop. Growers will have to assess and cost their individual storage needs. Growers who direct-market will often have to store a relatively small quantity of a wide variety of products. To offset this, cosmetic quality is not so important and produce might not need to be washed (and so reduce handling). Growers who market to packers or wholesalers might have large volumes of produce, but might only need to retain them for shorter periods of time. Some of the more commonly used storage methods are outlined below.

Field Storage: Storage of vegetables *in situ* in the field, generally over winter, is suitable for crops such as parsnips, swedes, carrots and savoy cabbage (until March). For parsnips and carrots, this provides the best storage method to preserve skin finish for the supermarket trade, where the carrots are sold washed. Carrots usually need insulating with straw to protect against frost and the large quantities needed (on a field scale) can make it costly and difficult to dispose of. However, field storage is, in general, low cost and simple to implement, although it is not always appropriate, for example on heavy land, where rotting may become a problem. Bad weather in winter can also make lifting impossible.

Clamps: Clamps (indoor and outdoor) are suitable for bulk storage of vegetable crops and are often used for root vegetables. They use ambient temperatures and ventilation and can be placed in adapted buildings such as barns. They are generally above ground and can be walled-in using hay bales. Some traditional clamps also used pits in the ground. Clamps are often covered with straw or other material to protect the produce, dampen-down temperature fluctuations and keep the produce in low light. They may or may not be ventilated by placing the produce on pallets and incorporating wire mesh. They are suitable for short (until December) to medium (until March) term over-winter storage but rodent damage can be a problem, as can frost damage if they are placed outside. All involve low fixed-capital investment, are relatively easy to construct and so allow flexibility in decisions on whether to store or to sell off the field. They result in low annual running-costs per tonne (£2–12/t) and are, thus, suitable for use with crops, such as onions, swedes and beetroot, for the wholesale and supermarket trade. Carrots and cabbage can also be stored this way for direct-marketing outlets. Potatoes will only store in clamps without sprouting until January/February. Storage beyond this point would only be satisfactory for the direct marketing of small quantities, where it is feasible to remove the sprouts prior to marketing. Presently, price increases over the winter season will adequately cover the cost of these forms of storage.

Box storage: Although it requires higher initial investment, box storage is often more suitable than bulk storage. It is the most practical way to keep different crops separate within the same store and any storage rots are likely to be kept localized within particular boxes. Boxes can be removed from the store as required and damage during handling can be minimized. Boxes can be tailored to the quantity of produce available for storage and can range from large pallet-boxes to crates and cardboard boxes of various types. Crops that are suitable for storage in clamps can also be stored in pallet boxes and more delicate produce in smaller crates or cardboard boxes.

Ambient air-cooled stores: Modified and (highly) insulated stores that have fan-assisted ventilation can provide relatively cheap (£25/t) and more reliable storage, especially for larger tonnages of most hard vegetables that require storage over the winter period. However, greater price premiums must be achieved to give a satisfactory return on the amount of capital invested. Produce is generally stored in boxes or sacks within stores, but in such a manner that airflow can be maintained around them.

Refrigerated storage: Refrigerated stores involve much increased investment and running costs. However, they can often maintain produce in better condition for longer periods than ambient-temperature stores. The use of refrigerated containers offers a reasonably low-cost alternative for an individual grower who has small quantities of produce and these can often be bought second hand or reconditioned. Refrigerated cold stores can allow rapid cooling of produce, which is important to maintain quality in perishable crops. A number of cooling technologies exist but are beyond the immediate scope of this book, although they include conventional refrigeration as well as moist-air cooling, hydro-cooling and vacuum cooling.

Plastic crates are a practical method of moving and storing produce

Once again, produce would generally be stored in boxes, crates or sacks within stores but, in this case, it is important that airflow be maintained around containers, so that they can quickly attain and remain at the low temperatures and in order to prevent moisture build-up around produce.

Long-term cold storage: Long-term cold storage is possibly the most expensive option (costing £30–40/t) and will require that a premium is obtained from selling produce off season to make it profitable. Economies of scale exist for larger stores, which, in the case of many small producers, point the way towards co-operative ventures in storage.

Controlled atmosphere storage: It is possible to control vegetable or fruit respiration by reducing the oxygen content of the air and increasing the carbon dioxide and/or nitrogen content. It is also possible to delay maturation and senescence by removing ethylene (itself generated by the stored produce or micro-organisms) from the air in-store. In the future, this developing technology will be useful for prolonged storage of several vegetable crops. Its use is already permitted in EU and IFOAM standards. The position over the permitted use of this technology needs to be considered and clarified by ACOS and other UK-approved organic-sector bodies.

Mixed storage: Organic growers tend to produce small quantities of a wide range of crops and find that they need to keep different crops in the same store. There is relatively little published information on long-term storage of different crops together. Most information is for transient storage of relatively perishable crops. However, mixed storage is possible and organic growers practise it with success for crops such as potatoes, carrots, onions and cabbage in refrigerated storage. It is also possible with the simpler forms of storage. Cross transfer of odours might be a problem in some situations and storage conditions will be a compromise between the conditions required for each individual crop.

ALTERNATIVE STORAGE AND PRESERVATION METHODS

Apart from bulk or cold-store storage, a range of other storage possibilities exists for organic vegetables. Most of them involve some form of processing before storage and so may need some form of investment to implement. They are likely to be more or less labour-intensive and probably also more suited to direct sales. The main alternative methods include freezing, drying and canning, and these are briefly outlined below. A detailed description of each is beyond the scope of this book, but information is relatively easy to find.

Freezing vegetables: Freezing builds on the principles of refrigerated storage and uses much lower temperatures (down to –20°C). It essentially arrests the processes of deterioration and decay. Vegetables should normally be blanched before freezing by putting them briefly into boiling water or steam and then reducing their temperature rapidly (say, in iced water). Foods will need to be packaged for storage in a freezer – popular choices include plastic bags and plastic containers – and should be sealed. Frozen foods can be sold in the packaging in which they are frozen, direct from freezing cabinets.

Drying vegetables: Removing the moisture from vegetables will also slow the processes of deterioration. However, it also alters the texture and the taste of the produce. Traditionally, vegetables have been dried in the sun, using various methods from simply spreading thin slices out in the sun to more advanced solar drying cabinets. More recently, gas or electric oven-drying has provided a more controllable alternative. Once again, vegetables would normally be blanched before the drying process and pasteurized afterwards. Drying techniques can remove 80–95 per cent of the moisture, at which point the vegetables will be brittle or crisp. Dried vegetables should be stored in suitable sealed containers (to protect them from insects and rodents). In a cool dark store they may last up to six months.

Canning vegetables: Pressure canning is the only safe method for home-canning vegetables. The bacteria *Clostridium botulinum*, which causes botulism food-poisoning, can develop in low-acid foods, such as vegetables. The bacterial spores will only be destroyed when the vegetables are processed at a sufficiently high temperature (116°C) for sufficient time. Canning, therefore, requires the use of specialist equipment and skilled labour and specialist advice should be sought.

Other processing methods: Many vegetables can be processed to make sauces, pickles, chutneys and relishes. Many recipes exist, but are outside the immediate scope of this book. They are generally available in a wide range of books. These processes generally involve either some form of pre-cooking before adding an agent that will preserve the vegetables or, alternatively, bottling cooked-sauces in jars under sterile conditions. Elevated levels of sugar or brine or immersion in pickling vinegar are usually sufficient to preserve vegetables. Vegetables prepared in this way can normally be stored for some time under cool and dark conditions.

Chapter 10

The Vegetable Market and Marketing

Production in every business should be geared to its market. Vegetables are no different and organic vegetables are a challenging and competitive market. Before planting, all vegetable growers should understand the real needs and demands of their customers or consumers by doing some market research, in order to get an understanding of the market potential and possibilities. This can be summed up as a process that:

- identifies opportunities and customers
- develops products that satisfy opportunities and customers at a suitable price
- presents and promotes the product to appeal to customers.

Thorough market research is more likely to lead to a successful and profitable business. In order to achieve this, the target customers must be identified, products developed, the market monitored, changes identified and alterations to products planned and presented as necessary. This chapter aims to help farmers and growers begin this process by providing an outline of the development of the UK organic-vegetable market and by briefly describing the various marketing outlets currently available to farmers and growers, together with an idea as to how to approach them.

DEVELOPMENT OF THE ORGANIC VEGETABLE MARKET IN THE UK

The organic food market increased rapidly between 1994 and 2002 but, since then, growth has slowed as the market has consolidated and matured. Nonetheless, organic food in the UK has a retail value of over £1 billion, which makes the UK market the third largest in the world. Vegetables are likely to be the first organic produce that consumers try and

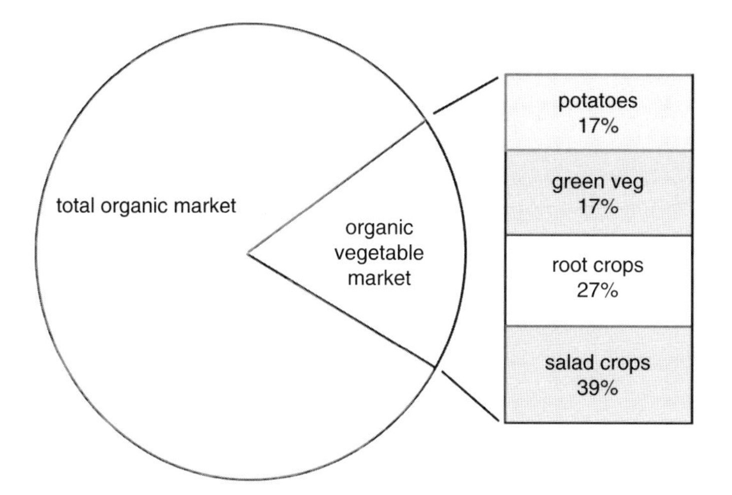

Figure 10.1 Organic vegetables as a proportion of the total organic market (2004)

which they return to time and time again, resulting in the vegetable market being the largest sector within the organic market, with a retail value of £200 m (18 per cent of the total in 2004). The market is split between potatoes, green vegetables, root crops, and salads, as illustrated in Fig. 10.1.

Since the late-1990s, supermarkets have been the primary outlet for organic vegetables, with sales increasing year-on-year. In recent years, the relative volume of organic vegetables sold through supermarkets has fallen marginally. On the other hand, direct sales of organic vegetables (through enterprises like farm shops, box schemes and farmers' markets) have increased rapidly, mainly as a result of the increase in number and size of box schemes. Percentages of organic vegetables sold through the four main outlets are shown in Fig. 10.2.

The level of imports of organic vegetables, although still high, has fallen to 41 per cent. Yet there are vast variations between different vegetables, from 70 per cent of sweetcorn imported to just 4 per cent of swedes. UK supply and imports for different crop categories are shown in Fig. 10.3, in terms of total tonnes sold.

The key characteristics of the organic vegetable market (as of the 2004 season) can be summed up as:

- the total market for organic vegetables was estimated at 123,500t, with a retail value of £197 m; of this, 73,500t (60 per cent) are grown in the UK

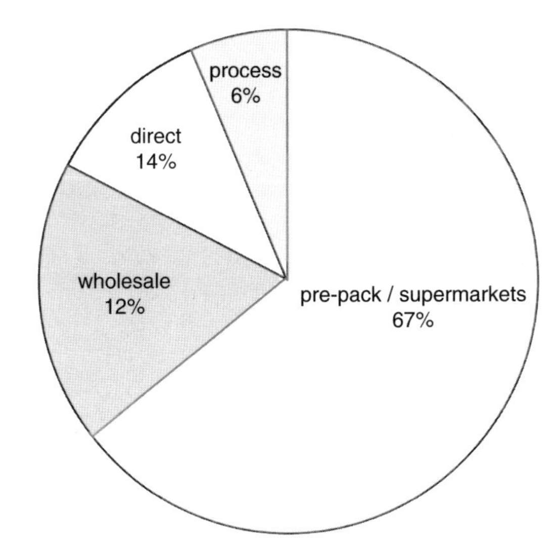

Figure 10.2 Percentage of organic sales by outlet (2004)

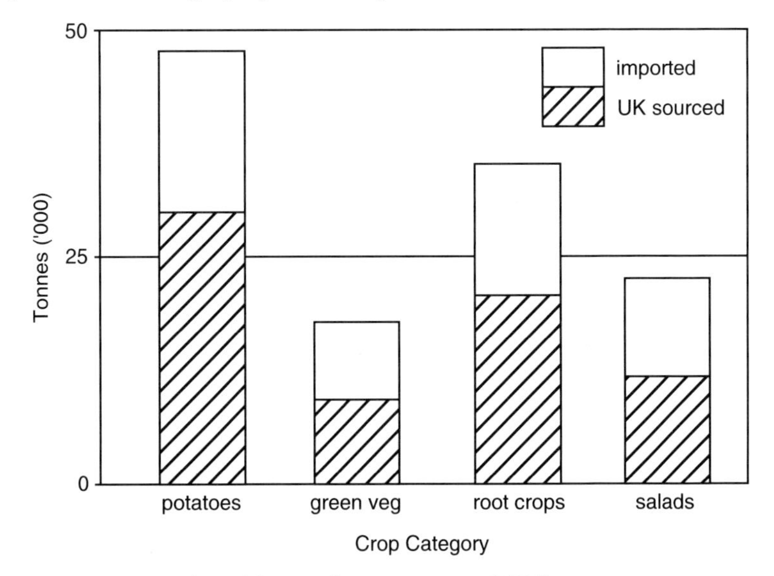

Figure 10.3 UK supply and imports by crop category (2004)

- 67 per cent of the vegetables are sold through pre-packers to multiple retailers (supermarkets), 12 per cent sold through wholesalers, 14 per cent directly by growers and 6 per cent sold to processing outlets
- the UK appears to have the highest quality specifications for organic vegetables in the EU

- growth in the organic-vegetable market is beginning to slow down compared to the initial growth period
- direct marketing is increasing and the relative volume sold through supermarkets is falling marginally
- in order to develop markets, UK growers will have to compete with imports (from within the EU and further afield) as well as other UK growers
- future growth in the sector will depend, in part, on further development of the infrastructure for processing and distribution and on novel outlets such as public procurement and catering
- there are many positive prospects for future growth of the market, including consumers' preference for UK and local produce, and opportunities to supply catering and institutional food-outlets with local food
- packers and wholesalers generally think there is potential to increase the market
- UK growers and marketers have responded to previous disorganization in the market by increasing specialization in specific crops or groups of crops, by buying each other out to reduce the numbers of players in the market and by focusing on continuity of supply
- the market is still vulnerable to fluctuations in supply. In the past few years there has been downward pressure on prices, especially for sales through supermarkets. Despite the upbeat editorial in the food trade-press about rising retail sales of organic and local produce, it is apparent the rising demand is not necessarily stimulating higher prices for farmers (probably due to imports).

THE DIFFERENT MARKET CHANNELS

There are many market routes available to farmers (*see* Fig. 10.4). The most suitable will depend to some extent on the resources available to, and the skills and desires of, the grower or farmer. First, we examine the general principles that apply to all the main outlets and provide a brief overview. Subsequent sections provide more information on the various marketing channels that are likely to be important to organic vegetable growers.

General Principles

Regardless of crop or intended outlet, it is essential that a market be secured prior to growing. Growing on speculation is both risky for the grower and, potentially, disrupts the market if large surpluses become available. Any marketing strategy must be researched and planned carefully to ensure it is

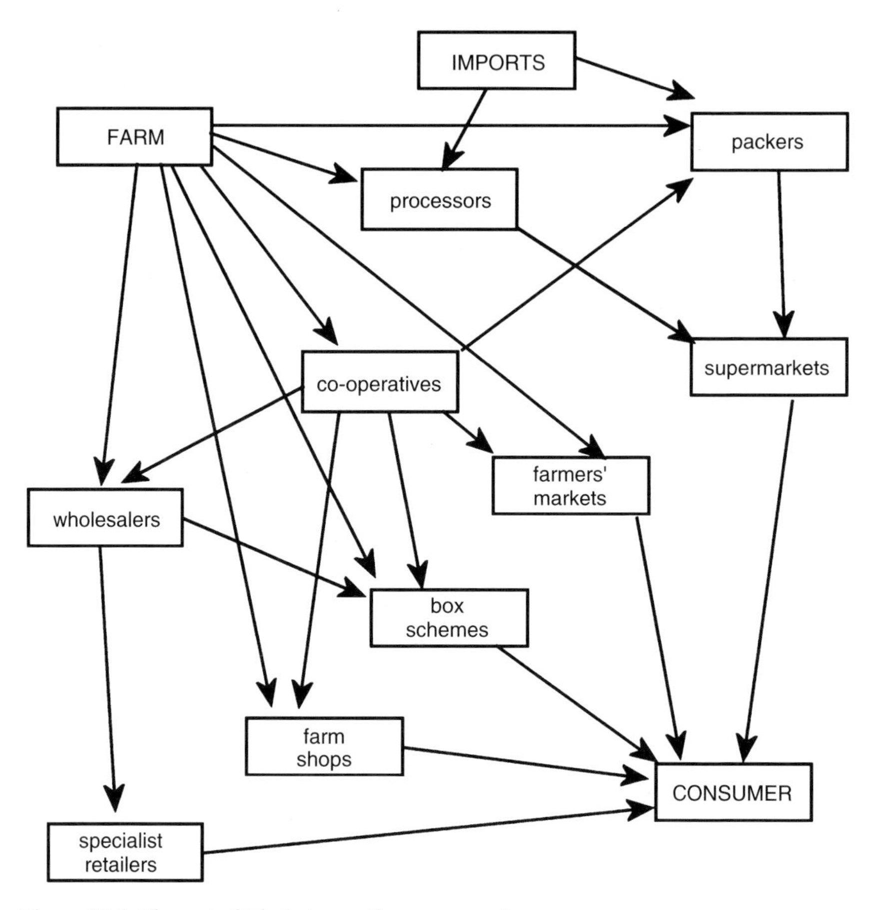

Figure 10.4 The main links between the grower and consumer

sustainable, feasible and practical. This involves talking to your potential customers, whether they are members of the public, packers or wholesalers, to gauge demand. It must be known who and where the customers are, what type of products they want and how much they will pay. A business plan, cash-flow forecasts and a production strategy should be formulated. Competition also needs to be assessed, sales volumes predicted and continuity of supply ensured.

One key point for organic marketing, regardless of outlet, is to win and keep the customers' confidence. People who buy organic produce are generally concerned about the environment and health. They therefore want to be sure they are getting genuine organic-produce and that the higher price is worth it. The marketer must, therefore, emphasize the continuous inspection process (normally backed up by a certification label) and the quality of their products. Additionally, the crop should be presented in a manner consistent with organic principles – minimal packaging, clear labelling and other desirable features.

Regardless of the outlet, the grower must know what is required – product specifications and supply chain practicalities – and how they will meet them. Each marketing channel has distinct features and requirements that are appropriate for different growers and production systems. Pack-house prices are generally lower than wholesale, but are an important outlet for large volumes. Direct outlets may obtain higher marketable yields and/or prices, but the added labour, packaging and transport costs must be carefully considered. The very wide range of crops that need to be grown may also restrict opportunities for mechanization and specialization. Table 10.1 highlights some of these market features.

The EC class system is a common, legal, Europe-wide marketing standard for each type of vegetable. Its purpose is to define the quality requirements, generally after preparation and packaging. Provisions cover quality, size and presentation. Class I requirements tolerate a lower level of defects and less variation within lots, or packs, than Class II requirements. Supermarkets also have 'in-house' class specifications that can exceed standards in the EC class system.

To make a success of marketing, the producer must also be cost aware (*see* Chapter 11). Operations must always be innovative, efficient and focused on high quality and growers must be ready to make improvements. One way of reducing capital costs is to keep infrastructures lean and purchase second-hand equipment. Advertising costs can be minimized by making full use of free marketing, such as parish magazines and leaflets. Key events in the area can also be used to promote the business. A website to promote the business has to be kept up to date and publicized and will cost £1,000 or more to set up, depending on the level of interaction provided for visitors. There will be small, but recurring, on-going annual costs to host and maintain the site.

It is important to remember that special consideration must be given to the VAT aspects of diversifying the farm business, as the new activity may not be treated the same as the existing farm business from a VAT standpoint.

DESCRIPTIONS OF MARKET OUTLETS

Packers and Supermarkets

Market Size: the first multiple retailer stocked organic food in 1981. Growth was slow in the early 1980s, but there has been rapid expansion in

Table 10.1 Summary of the features of the main marketing channels

	Pre-pack/retail	Processing	Wholesale	Direct
Marketable yield (% of total)	Lowest (50–70%)	Medium (70–90% as per contract)	Medium to high (70–85%)	High (90–95%)
Specification EC Class	Class I and II	—	Class II	Class I and II
Costs and prices	The grower receives a price, which is net of grading, packaging and transport costs. Some growing costs will be higher to meet the higher quality specifications, others will be lower due to economies of scale.	Highly capital-intensive and price-competitive, but capable of taking large volumes.	Growers need to pay all market charges (grading, packaging, transport and commission). Growing costs are high due to large number of crops grown. Inability to take advantage of economies of scale and mechanize operations.	Higher growing costs than wholesale. Marketing and management (planning and organization) expenses high. High labour costs to include delivery. Usually need own delivery vehicle.
Advantages	Ability to take large volumes means it suits large-scale operators. Less need for storage, grading or packaging facilities	Product grown to order at an agreed price.	Can cope with smaller quantities so more suited to small–medium sized operators. Specifications are less rigorous.	Highest prices received. More control over price. Suits smaller-sized growers. Close relationship with

Table 10.1 (continued)

	Pre-pack/retail	Processing	Wholesale	Direct
	on-farm and labour associated with these activities. Regular payment for vegetables.			customers. Lower levels of waste, generally less distance travelled. Produce fresher. Payment received when vegetables delivered.
Disadvantages	Less control of specifications and grade-out levels. No control over price and losses if specifications are not met. Remote from customer.	A contract must be secured prior to growing. Tight specifications. Losses if specifications are not met.	Standard payment conditions are 30+ days after delivery.	High labour requirements. Much effort, enthusiasm and time required to establish market.

both the volume of sales and the number of lines available since the late 1990s. In recent years supermarkets have increased their volume share of the organic market year-on-year, although recent seasons have seen a stabilization in their volume share of the market (currently around 67 per cent in 2004).

Market characteristics: Most of the major retailers have a very small, select, supplier base (the packers) that source organic and conventional crops from the UK and overseas. Generally, buyers talk to existing packers rather than the growers. Most multiples have been improving communications with their supplier base in recent years and have organized producer groups and discussion, or technical, days, as well as farm visits. If producers are unsure who is supplying the major multiples, they can, theoretically, contact a retail buyer directly and negotiate with them.

Packers have very tight specifications on quality, variety, taste, size and colour, which are normally set by the supermarkets they supply. Consistency of quality and supply is very important, as repeat purchases can make or break a line in supermarkets. Pack-house prices are generally

Supermarket sales dominate the organic vegetable market

lower than wholesale prices, but they are often the only realistic outlet for large volumes of produce. Grading is often carried out at harvest, on the farm and at the pack house.

Hence for any major retailer the key issues are the same:

- continuity of supply and quality
- pricing in line with competitors
- quality (appearance and taste).

Points to Consider:

- Can produce that does not meet requirements of multiples be marketed (as major multiples only tend to take the premium quality produce)?
- Organic produce must legally be separated at all stages of the marketing chain, so can you ensure no cross-contamination can happen?
- Can you produce the quantity, quality and consistency to meet retail requirements?
- Can you cooperate with established suppliers or find a unique selling point?
- Don't expect to go it alone unless you have sufficient scale of operation.

Table 10.2 Benefits and constraints to supplying packers and supermarkets

Benefits	Constraints
Require less on-farm storage, packaging and grading and the associated costs (in most circumstances)	High quality requirements therefore high growing and harvesting costs and a lower marketable yield than other outlets
Few crops need to be grown so crops can benefit from economies of scale	There may only be demand for some of the crops in the rotation
High volumes can be marketed	High continuity demanded

Wholesale

Market size: About 12 per cent of organic vegetables are sold through wholesalers. Where major supermarkets largely rely on direct contacts with packers, independent retailers source many of their products through wholesalers. As the number of independent retailers and box schemes grows, so does the need for organic wholesalers. However, as middlemen, they are dependent on the prosperity of their customers and require a share of the retail value.

Market Characteristics: in comparison to packers, wholesale outlets generally take lower volumes and quality, thus allowing a higher marketable yield. However, costs may be higher due to the need to pay on-farm packing and grading, and for transport and marketing. Generally, a larger range of crops also needs to be grown to supply wholesalers. Organic-vegetable wholesalers rarely work out of the traditional markets, since they don't require the facilities the traditional markets offer. They are largely located close to growers (for instance, in areas of traditional small-scale growing like Herefordshire) or close to consumers in large cities.

For the wholesale sector, as with other retail outlets, quality, consistency of supply and price are the key issues.

Points to Consider:

- Can you develop links with wholesalers who demand the crops you can produce?
- Can you meet the specifications and terms (say packaging or transport) of the wholesalers?
- Can you ensure no cross-contamination will occur with conventional vegetables, and provide the relevant documentation?
- Can you produce the quantity and quality demanded?

Table 10.3 Benefits and constraints to supplying wholesalers

Benefits	Constraints
Smaller quantities required (therefore suits smaller and medium-sized producers)	Supplying the right quality for the right price
Specifications more flexible than for the pre-pack market	Higher transport costs than marketing directly to the consumer
Most crops in the rotation are likely to be marketable	On-farm packaging may be required

Direct Sales

Market size: As a percentage of total market volume, direct sales of organic vegetables have slightly increased in recent seasons. This is a result of strong growth in both the number and size of box schemes and delivery services (a 16 per cent increase in 2003) together with an increase in sales through farm shops (by about 10 per cent).

Market characteristics: there are various outlets where growers market direct to consumers thus excluding middlemen prices and specifications. However, this means the grower has to take responsibility for quality, logistics, marketing and any legal requirements, on top of any growing that is required. Some general information about direct marketing is provided below, followed by more detailed information for specific outlets including farm shops, farmers' markets, box schemes and community supported agricultural schemes (CSAs).

General Characteristics of Direct Sales

The nature of direct sales means that it is necessary to build a solid relationship with customers. Market research, as with all outlets, is essential. Yellow Pages, local papers and phoning potential competitors posing as a customer can all be invaluable, yet simple, methods of determining local competition and demand. Direct sales generally require that the grower (and/or family members and employees) be committed to retailing and to providing consistent and reliable service, as well as projecting a good image. One method of developing a relationship with customers is to send out regular newsletters, with a questionnaire to get feedback. Farm walks, open days, weeding or harvesting days offer further opportunities for creating interest and a sense of community with customers.

Quality and freshness are of utmost importance for all direct-sale outlets, as they are strong selling points. The company must ensure production and harvest techniques are appropriate to ensure this (*see also* Chapters 7 and 9). Additionally, the business should be capable of meeting consumer expectations; this might include being open at weekends and in the evenings and offering a wide range of products.

Through dealing direct with customers the business will be responsible for setting its own pricing and payment policies and keeping track of payments. The net return must be a figure that represents an acceptable level of earnings. It is common practice to use supermarket prices as a guide, but care should be taken not to undersell the business. Business Link is a good source of information on price setting and effective selling (http://www.businesslink.gov.uk). The long-term success of direct sales depends on continuous development of the business and keeping costs under constant review. It is not sufficient to maintain systems that are working; continuous development and a proactive approach are essential for the future.

Resources Required

The nature of direct selling implies that the farmer or grower needs to deploy a wide range of resources. The most important of these are mentioned below.

Land: the length and variety of cropping required for most direct sales puts high demand on land and so it needs to be good quality and highly productive. A water source for irrigation is also advantageous. Remember that currently extraction from watercourses requires an abstraction licence from the Environment Agency. Car-parking facilities and good access to the farm will also be necessary, if you intend customers to come to the farm.

Buildings: Shop, packing and/or storage areas will be necessary, depending on the selling method. Planning permission is required to convert an existing building if the use is to be changed or products are brought in to resell. It may be necessary to erect polytunnels, to ensure continuity of supply (*see* Chapter 8).

Packing area: ideally this should allow the packing units to be laid out, side-by-side, so quality can be monitored and they can be lifted out easily. Space may also be required for weighing and bagging.

Storage area: for crates of vegetables coming in from the field. The area should be shaded and cool. Long-term storage may also be necessary for crops that have a single, main harvest. These areas must be rodent proof, cool and frost resistant. Shipping containers can be a cheap starting point (*see also* Chapter 9).

Cool store: this is desirable but not essential for all systems.

Shop area: this will depend on resources available, but will form an impression of the business for customers.

An attractive shop area is necessary to form a good impression

Visitor facilities: if customers are going to visit the farm they must be accommodated. In some circumstances they may also need toilet facilities.

Vehicle: if customers do not come to the farm, produce will need to be taken to them. A car and trailer may be sufficient and will reduce capital expenditure. The business name and telephone number on the side of the vehicle can be effective advertising.

Machinery and equipment: specialist machinery and equipment might be necessary, in some cases, to move produce around, prepare it or display it. Tills and other shop equipment might also be necessary.

Labour: harvesting and preparation time will, hopefully, be consistent throughout the year, although during the growing season a considerable amount of time is needed for land preparation, sowing, weeding and general crop care, hence casual labour may be necessary. Training of staff may also be necessary. There is a variety of retailing and management courses available – for example, local agricultural colleges or technical colleges.

Customer base: this can be an initial challenge. Publicity will be necessary to develop and attract customers. This can be in the form of attending farmers' markets, developing links with local groups or advertising in shop windows, parish magazines or local newspapers. Signposting to the farm will also alert customers to your business; however, this may require planning permission, or agreement from local landowners.

Regulations

Facilities where fresh produce is handled should be registered with the local health environment office and a visit can be requested to sort out any problems. The Food Standards Agency can help with general food-hygiene issues. Businesses are subject to all trading-standards legislations and must expect an unannounced visit from an inspector. The acts that cover this include the Food & Hygiene Acts and Weights & Measures Act. Shops should also ensure that they are complying with any organic standards laid down by their certifying body. Normally they would expect to be covered by organic processing and retail standards.

Insurance is now essential for any business that employs labour or deals with the public. Public liability, product liability, employer's liability and vehicle and premises cover for theft and damage should all be considered. Local NFU offices can assist with employment information.

Investment Costs

In providing the material resources required and meeting the legal regulations, some form of monetary investment will need to be made. Whilst costing information varies depending on the exact circumstances, £35,000 would probably be a safe minimum estimate for setting-up a permanent retail outlet. A high turnover is needed to repay the investments and costs but will vary considerably, depending on size of outlet, products on offer, length of season and other factors. Research by Cardiff University for the Farm Retail Association found that the median (the most common) farm-shop turnover is in the region of £90,000, providing a median net income of £12,000 per year.

Specific Direct-Sales Outlets

Farm shops: Increasingly, consumers are interested in how food is produced and what it contains – farm shops are ideally placed to meet this demand. Farm shops can range from a temporary stall selling seasonal products to a full-scale, full-time 'mini-supermarket'. The produce sold will depend on the type of farm, but usually includes a range of fruit and vegetables, fresh meat, dairy products, bakery products and jams, to name but a few. For this type of marketing to be successful, the farm must be easily accessible. Ideally, it should be located near consumers (villages or towns) or close to a busy road. The SA Organic Directory lists 337 farm shops, yet there are likely to be many more, including non-organic farm shops that sell some organic lines.

Points to Consider:

- How accessible is the farm?
- How will you attract customers and promote the business?
- Can you take advantage of seasonal markets, such as Christmas trees, soft fruit?
- Can you run a box scheme as well (*see* below)?
- How strong is your brand?
- Will you sell your own produce and/or buy in? Where and who will you buy from?
- Is it worthwhile buying some crops that are available cheaper on the wholesale market than you can produce them? Will this affect consumer demand?
- Can you work collaboratively with other local growers? If you are considering this, how will income be apportioned?
- How can you overcome the 'hungry gap' (the time when local fresh produce is generally difficult to source) between March and June?
- Can all produce reaching the consumer be fresh and harvested according to demand?
- How will customers pay for produce? Are debit- or credit-card facilities required?
- What competition will have to be faced from other retail outlets in the locality?

Table 10.4 Benefits and constraints to running a farm shop

Benefits	Constraints
Customers come to the farm so distribution issues are removed	A wide range of crops should be grown so limiting economies of scale
Transport and packaging requirements are less, thus reducing the producers' costs	Farm must be accessible
Increased financial returns through direct selling and improved cash-flow	Safety and security issues of customers being on the farm
Direct customer feedback on produce and prices	Sometimes antisocial and long hours
A secure and regular market outlet can be developed	Extensive investment may be necessary

Farmers' markets: The first farmers' market in the UK was held in Bath in 1997. Since then numbers have risen to 500 markets countrywide, with a concentration in the south-west and south-east and a dedicated organic market in north London. Farmers' markets provide local small-scale producers and processors a forum in which to sell direct to the consumer and contribute to the local economy. For the consumer, farmers' markets offer direct contact with producers, thus traceability, and normally a more relaxed shopping experience. Producers, in turn, get valuable feedback from their customers. With the increase in market numbers, it is possible for individual producers to attend a substantial number of different markets.

The National Association of Farmers' Markets (NAFM) was established in 2002 to monitor and guide the conduct of these markets. In order to be a member of the NAFM the market must adhere to certain criteria. The main factors that affect producers are:

- Locally produced – only goods produced in a defined radius of the market, normally 30 miles (50km), can be sold.
- Own produce – all produce sold must be grown, reared, caught, brewed, pickled, baked, smoked or processed by the stallholder.
- Principal producer – the stall must be attended by the principal producer or a representative directly involved in the production process.

Points to Consider:

- Is there a farmers' market operating in your area (the Soil Association produces a list of current farmers' markets operating in the UK)? How frequently is it held?

Table 10.5 Benefits and constraints to working a farmers' market

Benefits	**Constraints**
Good public relations and advertising for the business	Potential time-conflicts when work needs to be done on the farm
A secure and regular market outlet	Will only access a limited customer base
Increased financial returns and improved cash-flow through direct selling	Weather dependant
The producer can get direct customer feedback on produce and prices	Other outlets may be necessary to sustain the business
Transport and packaging requirements are less, thus reducing producers' costs	Local councillors can make it difficult to hold markets in good locations and in-fighting amongst stallholders can be a problem
	In some areas they do not operate regularly

- Can you contact your local authority to ask if they have plans to establish one?
- How can you transport goods and display them at the market?
- Can you generate enough income through farmers' markets to sustain the business or do you need additional outlets?

Box schemes: Box schemes are one of the fastest growing methods of direct marketing for organic growers. There are several hundred now operating in the UK, ranging from supplying less than 100 boxes a week to supplying many thousands. As the total volume sold through box schemes is still small, but increasing, there are substantial opportunities for expansion. Currently there are about 500 businesses, with sales of almost £42 million, according to the Soil Association. As a benchmark, about 100 boxes a week requires one full-time worker and a part-time assistant. About sixty boxes can be produced per hectare.

Box schemes have proved to be a simple and effective method of marketing and distributing organic vegetables for many growers. Whether they are farm-based (where a grower markets most of their own produce) or non-farm-based (where they often operate like wholesalers) the success of a scheme lies in offering the best possible service, to find and keep customers happy, and the ability to compete with large retailers. The key aspects of this are variety, quality, overall value and a close relationship between the consumer and producer. Typically, vegetables are the main constituents, although fruit, meat and dairy products are also often included. The aim is to deliver high quality, affordable, organic produce to the consumer, with a reasonable financial return to the grower.

Most box schemes operate at a local level, but several have developed nationally and offer franchises. For producers, this means there are several options: starting up your own scheme, supplying another scheme or taking on a franchise. The quantity and variety of produce needed will vary with these options. Table 10.6 shows the typical content of summer and winter boxes sold at two different prices.

Points to Consider:

- Will you deliver or will customers collect their own boxes? If you deliver do you need a series of drop-off points?
- How often and when will you pack and deliver?
- What type of boxes, or bags, could you use?
- Is it possible to reuse the boxes and so reduce cost and waste?
- Can you produce enough to satisfy the customer's full needs until the next delivery or will you need to buy-in some produce?
- Do you have the skills and resources to produce a wide range of vegetables through the majority of the year?
- Can you develop a covered area for packing?
- Can you handle the logistic side of managing the boxes, addresses and consumer preferences – among other factors?

Table 10.6 Examples of summer and winter medium-sized boxes

Box sold at £5.25		**Box sold at £8.00**	
Summer	*Winter*	*Summer*	*Winter*
1–1.5kg potatoes	1–1.5kg potatoes	1.5kg potatoes	1.5kg potatoes
0.5kg carrots	0.5kg carrots	200g mushrooms	350g red onions
Lettuce	0.5kg onions	350g carrots	450g carrots
0.5kg beans	Cabbage	450g broad beans	350g parsnips
0.5kg tomatoes	Salad pack	350g calabrese	500g swede
1 bunch beetroot	0.5kg parsnip	1 bunch turnips	700g leeks
0.5kg courgettes	0.5kg Brussel sprouts	Cucumber	450g Brussel sprouts
	0.5kg leeks	Celery	350g pak choi
		Kohl rabi	1kg squash
		1 bunch beetroot	1 bunch parsley
		450g spinach	
		1 lettuce	

Direct sales through box schemes are growing in popularity

Table 10.7 Benefits and constraints to running a box scheme

Benefits	Constraints
Loyal customer base and secure income can be developed	Complexity of cropping and management to supply a wide variety of vegetables throughout the year
High profitability off small areas of land, which do not have to be accessible to consumers	Cannot benefit from economies of scale from growing a large area of a single crop
Direct contact with the customer	Storage facilities may be required for main-crop vegetables
Generally the grower has full control over the business and the produce which is sold	A well-organized system for managing payments and deliveries to many individual customers is necessary
Can form a market for out-grades from other outlets	During peak harvest-time other outlets for gluts of vegetables may be necessary
Minimal packaging costs are possible	Managing boxes and consumer preferences logistically challenging
Allows experimentation with novel crops and varieties	
Financial risk is spread over a large number of customers and crops	

Community Supported Agriculture (CSA): CSA is based on members holding shares in the farm business and getting a weekly vegetable-box in return. The farmer and members decide what the farm can produce and what they would like to receive, before the farmer develops a crop plan and budget for the season. This incorporates all production costs and a fair wage to the farmer. The budget is then put to the members for approval and the cost of an annual share is calculated by dividing the total cost between participating members. Hence the farmers and consumer share the responsibilities and risks of farming. Members may also take part in regular meetings and workdays. Although CSA has been adopted widely in the USA, it is still rare in the UK. The Cultivating Communities project, funded by the Lottery Community Fund and carried out by the Soil Association, was launched in 2002–03 and has raised the profile of CSA, although their definition of CSAs has been quite broad.

Points to Consider:

- What opportunities and demands are there in your locality?
- Who will be the stakeholders?
- Who has ultimate control of business decisions and how will the enterprise be managed?
- How will the scheme get working and start-up capital?
- How can you connect members with the farm and keep them informed?
- How will you develop a CSA?
- What input is required from all participating members?
- What could make it fail?
- What happens to assets if the scheme is wound up?

Table 10.8 Benefits and constraints to setting up a CSA scheme

Benefits	Constraints
Mutual support for farmer and consumers	Hard work for the organizers
In theory, a secure and fair return to the farmer	Modest income
Demand and potential supply are matched before production begins	Dependent on locating a committed membership base
Educates consumers about farming	

Processing

Market Size: There is growing demand for root crops and vegetables grown on contract to supply processors for prepared and frozen foods and baby foods (for example, organic baby-food now commands more than

50 per cent of the market). However, the growth of UK supplies for this market has been slower than expected and prices are often low. Many larger manufacturers source produce directly from a small number of suppliers and it is a highly competitive market.

Market characteristics: Processing can range from those repacking organic products to businesses adding value through extensive food-manufacturing. This section deals mainly with food manufacturing. Processing is, by and large, not suitable for grade-outs from other outlets, as the quality requirements are very specific. In some cases, growing for processing may require access to freezing and/or storage facilities, if the processor is unwilling or unable to take on a full year's supply at one time.

Vegetables can be processed in many different ways, depending on the type of raw material and the end product. The techniques most frequently used are canning or bottling accompanied by heat treatment, refrigeration or freezing, fermentation, drying and pickling. In most cases the aim is to lengthen the shelf life (reduce the perishability) of the product, but there are often secondary objectives, such as to make the product more convenient to use, to improve the packaging and presentation, to improve the eating quality or to produce an entirely new product, such as juices, purées, jams or wine.

The manufacturing steps include some or all of the following: receipt and weighing of raw materials, storage, washing, grading, peeling, cutting, crushing, filtration, heating, cooling, preservation, pickling, drying, concentration, fermentation, packaging (cans, jars, vacuum packs, tetrapaks and so on) and storage of finished products.

Typical processed organic-vegetable products include: baby food, canned sweetcorn, beans and the like, frozen peas, concentrated tomato paste, vegetable soup, potato chips (french fries), crisps and so on. The Processed Vegetable Growers' Association Ltd can be contacted for more information.

Points to Consider:

- Have you got a contract to supply a processor?
- Can you develop the necessary storage facilities (*see* Chapter 9)?
- Is there sufficient demand for you to develop a processing business on site? If so, large volumes of water are needed for washing raw materials, factory cleaning purposes and so on, so can you meet high water requirements and dispose of grey water? Additionally, processing can produce large volumes of bulky, perishable, solid waste, such as peels, stems, shells, rinds, pulps, seeds, pods, reject raw material and the like. How will you deal with this waste?
- Have you got the capital to meet stringent sanitary and environmental regulations?

Table 10.9 Benefits and constraints to supplying the processing industry

Benefits	Constraints
Scope for development of UK supplies if quality and continuity requirements can be met	Many processors are heavily reliant on imports
Expanding consumer demand for convenience and processed foods	Access to large population centre or processing plant will be required in some circumstances

Co-Operatives

Market size: Historically, cooperation among UK farmers has been low. Arguably the reasons for this are lack of opportunities, unclear benefits and a desire to remain independent. However, there is currently much policy support for developing agricultural co-operatives. Proven benefits of co-ops include effective understanding of consumer demands, delivering economies of scale, developing new products and increasing supply-chain efficiencies.

Market characteristics: The English Food and Farming Partnership (EFFP) is currently aiming to strengthen the profitability, competitiveness and sustainability of England's farming industry through the growth of market-focused farmer-controlled businesses (FCBs) and by developing cooperation and partnership activities between farmers, and between farmers and the food chain. Co-ops clearly fall within this remit.

Evidence from existing co-operatives suggests they must be backed by a great deal of planning and a clear strategy, with good corporate governance. Equally important is the commitment of executive staff, with the ability and freedom to manage the business on a day-to-day basis. Additionally, successful co-ops continually develop the business with a focus on sustainable and long-term benefits for members and for meeting customer needs.

Co-ops have the benefit of bringing together the experience and knowledge of all members for the advantage of the group. The English Food and Farming Partnership can direct growers towards additional suitable sources of advice and support

Points to Consider:

- Who can you collaborate with (look both vertically and horizontally along the food chain)?
- Will co-operations take the form of purchasing collaboration or marketing collaboration?

Table 10.10 Benefits and constraints to joining a co-op

Benefits	Constraints
Co-ops can help growers do things better than they would alone – for instance, more effective investment, reduced costs, shared skills and advice, higher output, continuity of supply and better prices	Co-ops can fail due to not being open and transparent and not operating to the highest standards to ensure the confidence of their members. They can also fail if members lack loyalty and commitment
They can enable farmers to look beyond the farm gate and to strengthen their market orientation	English producers lag behind competitors in collaboration
Large customers demand fewer, well-organized suppliers	Require substantial capital to set up
They can strengthen farmers' position in the supply chain and potential to access grants	Farmers may feel a loss of independence

- Will collaboration improve the long-term profitability and sustainability of your business?
- Will professional management be required (good management and governance is essential)?
- What function does the co-op need to fulfil (supply, marketing, processing)?
- Are there existing co-ops you could join?

Public Procurement and Out-of-Home Eating Opportunities

Market size: Public procurement is in its infancy, despite apparent policy support encouraging public bodies to benefit both local economies and the health and environment of local communities by procuring local food. This, combined with the government's support for healthy eating (for example five-a-day initiative), provides many opportunities to work on both a large and local scale. However, these government policies will have to gain far more support from public-sector institutions before growth occurs.

Market characteristics: The specific contracts through which public-sector institutions – schools, hospitals, local authorities – buy supplies is known as public procurement. This provides an opportunity for growers to supply large volumes of produce to these bodies, although there are some complex and challenging practicalities. Sometimes small local volumes are also needed. The contractor responsible for supplying a public authority has to be able to meet clear criteria. These include continuity of supply and consistency in quality, hence alternative markets for the proportion of produce that does not meet this standard may have to be found. Additionally, accountability is important, so reliable data on nutrition and allergy risks must be able to be provided, as well as full production history, storage instructions and quality-control information.

Public bodies, such as councils, schools and hospitals, will have a range of different procurement methods and there are a large number of procurement rules and laws, ostensibly there to protect taxpayers and prevent corruption. In general, specialist advice will need to be sought to approach these markets. In general it should be remembered that:

- contract opportunities valued below £25,000 might not be advertised
- contracts exceeding £25,000 in value will normally be advertised in the local press and in at least one national newspaper or a suitable trade journal
- various evaluation criteria can be used to judge tenders. However, the usual criterion is 'MEAT' (Most Economically Advantageous Tender), which allows the bodies to take other factors, such as quality, into account when making award decisions
- EU law makes it illegal for public bodies to specify that products must be sourced locally, although locality can be specified in other ways, such as freshness.

So-called out-of-home eating opportunities include outlets such as restaurants, bed and breakfast, private schools and lunchbox menus for children. Much disposable income is now spent in institutions like these and the long-term potential for developing these markets is promising but unclear. The market may, for instance, incur high costs in preparation, distribution and ensuring consistency, although it can be an important selling point in some circumstances, such as restaurants and pubs. Specialist markets, such as heritage vegetable varieties, might also be attractive within this type of marketing scheme.

Points to Consider:

- How can you connect with the various institutions?
- What products do they demand? Can you supply these in sufficient volumes?
- Is your council encouraging public procurement of organic vegetables?
- Can you meet the quality, quantity and continuity specifications?

Table 10.11 Benefits and constraints to supplying public bodies

Benefits	Constraints
Large potential market	A range of crops is often desired from one supplier
Can be profitable	Supplying schools will have lower demand in the holidays, but generally all-year supply is desired
Secure market	Obtaining contracts can be difficult

MARKETING ADVICE AND GRANTS

Marketing is a complex area and subject to many government policy initiatives. The government does, however, provide a range of advice and financial assistance to farmers and rural businesses to help them meet these policy initiatives. Whilst a detailed description of the assistance available is beyond the scope of this book, and anyway dependent on regions and subject to change, some of the schemes that could assist with the marketing of organic vegetables are sketched out below in order to give an idea of where help might be obtained.

Central Government (DEFRA / EU)

The England Rural Development Programme (ERDP) and its equivalents in other regions (Scotland, Wales, Northern Ireland) promote rural development through various schemes. These include the Rural Enterprise Scheme (RES), Processing and Marketing Grants (PMG), Vocational Training Schemes and the Organic Farming Scheme (OFS), Farm Business Advice Service and LEADER+. The relevant DEFRA Rural Development Service will provide information on all these and other schemes, although the Mid-Term Reform of the Common Agricultural Policy may affect some of them.

Local Agencies

Local Development Agencies: may have locally applicable sources of grants and advice to help establish small businesses and encourage the growth of existing businesses. They can advise on topics such as new business start-ups, business planning, financial planning and the day-to-day running of businesses. They can often arrange for a free legal consultation (45 minutes).

Local Authorities: may provide discretionary grants and low-interest loans, depending on individual circumstances.

Regional Agricultural Development Services: some regions also have these services, which provide free advice.

FUTURE PERSPECTIVES AND MARKET TRENDS

As the market further matures, and the increase in demand slows down (predicted at 10 per cent growth for 2003–04), especially in supermarket trade, the challenge to growers is to open up new markets such as public procurement and catering. Future seasons will see changes in subsidy payments and adjustments due to EU enlargement. Demographic trends and lifestyles are changing customer demand, and this will require the food chain to alter accordingly. Additionally, the effect that agricultural reform across the EU will have on incomes must be considered, as it is unlikely to provide the same level or type of support received in the past. A summary of the strengths, weaknesses, opportunities and constraints to increasing the UK market for organic vegetables is provided in Table 10.12.

Table 10.12 Summary of strengths, weaknesses, opportunities and threats in the UK market for organic vegetables (2005)

	Strengths	**Weaknesses**
Demand	Consumers choose organic vegetables for heath and taste reasons 70 per cent of consumers who have a preference for local organic food are willing to pay more for it Organic vegetables are a key entry point for consumers into the organic market Local origin is a key motivator	Consumer understanding of organic and local is often confused and intertwined Lack of consumer awareness and knowledge about what 'organic' means in terms of production system and certification Lack of irrefutable evidence to justify reasons for the organic claim to be better
Supply chain	The total market size increased 20 per cent by volume between 2002 and 2003; this increase in production is largely from existing growers	Deficiency of new and alternative high-volume markets and the necessary infrastructure to supply them

Table 10.12 (continued)

	Strengths	**Weaknesses**
	The proportion of crops sourced from the UK has increased slightly by 1.4 per cent (2003–04); UK continuity has also increased Relatively, direct-sales market-share is rising and the super-market share is falling Favourable euro/pound exchange rate during the last two years	Severe price competition between the major multiples High quality specifications for the UK market and conformance to EC grading standards Lack of production continuity can limit availability of UK produce Limited shelf space in super-markets can restrict availability
Supply	High level of grower expertise meeting demands of market specifications Better balance between supply and demand, less speculative growing and better quality Increasing optimism in the sector Skills base and expertise of organic growers increasing	Lack of a tendency for producers to cooperate Cost and uncertainty of conversion Some UK producers are not as professional as EU counter-parts Some growers are still producing before securing a market outlet
	Opportunities	**Threats**
Demand	To further develop consumer awareness and understand-ing of organics and food issues, in all sectors of society Promote health and taste as value for money Increase the focus on the link between food health and nutrition An increasingly health-conscious male population, which is a strong motivator for buying organic vegetables	Consumers don't buy organic vegetables due to high prices, confusion and mis-trust about labelling and cer-tification and low awareness of surrounding issues Consumers buy locally to sup-port the local economy and environment (health and taste are less apparent as drivers for local purchasing), hence they might tend buy local in preference to organic produce

Table 10.12 (continued)

	Opportunities	**Threats**
Supply chain	Develop security and communication in buyer–supplier relationships Develop local, regional and public procurement supply-chains and their infrastructure Further build direct-sales to fulfil consumer demands for local vegetables	Blurring of market boundaries as conventional producers introduce practices closer to organic standards Mismatch between producer returns and retail prices Uncertainty about the impacts of CAP reform and EU enlargement
Supply	Develop producer cooperation Utilize promotions from the Five-a-Day campaign Focus on producing high quality vegetables that meet consumer criteria for health, taste and environment benefits Enhance policy support Further work on breeding and developing the best varieties for organic vegetable production	Declining returns to producers threatening business sustainability Packer specifications leading to high waste-levels Rationalization and specialization may mean some packers and growers lose out

Chapter 11

Farm Economics and Business Planning

This chapter outlines the basics of analysing the economics of organic vegetable growing. It introduces key terms and outlines the advantages and disadvantages of different economic farm analysis methods and identifies factors that should be taken into account when planning for and running a sustainable vegetable production unit, or a mixed system with vegetables.

Many organic (and conventional) vegetable production units have found profitability increasingly difficult, due to factors beyond their control, such as increased competition as the organic market is maturing (*see* Chapter 10). This has lead to downward pressure on prices and lower profits. Hence, economic analysis of the business is even more vital, as it is only on this basis that judgements can be made about the performance of the business and/or enterprises that comprise it. In addition to increased overall efficiency, sound financial management also aids the development of more profitable enterprises, within a holistic whole-farm approach.

BASIC FARM ECONOMICS

The economics or profitability of a farm business is mainly determined by the income from marketable yield (output) minus the costs of production (inputs).

As shown in Fig. 11.1, inputs or resources used on the farms are items like land, labour and capital and their respective costs. Outputs are marketable yield multiplied by prices, plus any subsidies or grants that may be available, for example for environmental features. Several outputs can be achieved with different combinations of inputs, as some inputs are substitutes for others.

The economics of production is about making the best use or optimizing the use of the available resources (or inputs, factors of production) to maximize financial outcome and, at the same time, meeting the constraints, such as rotation and other technical or environmental requirements. Once

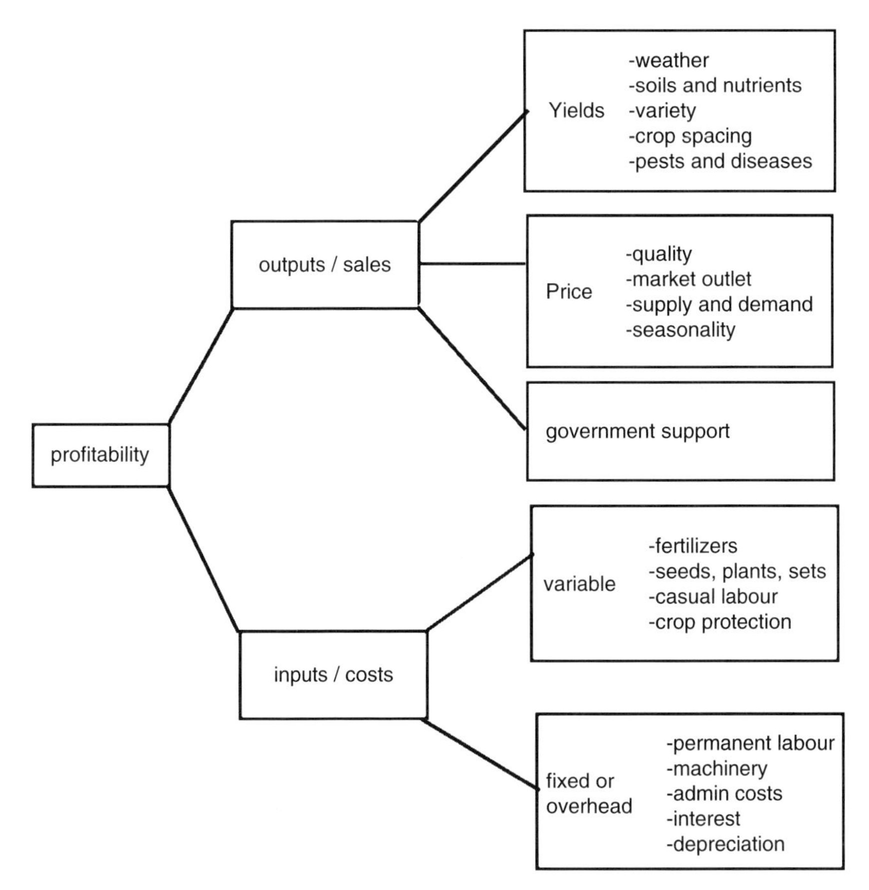

Figure 11.1 Factors affecting the profitability of organic vegetable production

the farm is understood in this way, it is possible to plan and match the farm performance to desired outcomes. However, whilst some factors are more or less under the control of the grower, or *internal* to the production system, some are definitely not. A brief introduction to some of the more important concepts is provided below.

Enterprises: Any farm is usually made up of a number of enterprises, such as individual crops (potatoes, carrots) or livestock (cows, sheep).

Costs: Costs, or inputs, can be split between variable costs and fixed costs as defined below.

- *Variable costs* usually change in direct proportion to the size of the enterprise – the number of hectares or head of stock. The main items are seed, fertilizer, and casual labour specific to the crop.

- *Allocated fixed costs*, such as costs of machinery operations (say, mechanical weeding), are not variable in direct proportion to the amount of production, but it is possible to allocate them to a certain crop enterprise.
- *Overheads or fixed costs* tend to remain the same regardless of how much is produced, for example rent, depreciation, interest and insurance.

Subtracting the variable costs from the financial output results in the *gross margin*. Similarly, subtracting the allocated fixed costs from gross margin results in *net margin*. Gross or net margins can be calculated on an enterprise level, on a rotational level or on a farm level. A sharp distinction between fixed and variable costs is not always possible, nor realistic, under farm conditions.

An economic analysis at the farm level allocates these inputs and analyses them to indicate where income can be increased by optimizing inputs. However, it should be borne in mind that it is often difficult to allocate inputs to all farm operations and that not all financially effective solutions will also necessarily be technically, agronomically or ecologically effective. They may also fail to meet market demands, as well.

Internal factors: are those that the grower retains some control over and can, to some extent, adjust. In the case of organic vegetables, the most important is the use of labour, which usually amounts to a high proportion of the total farm costs. Other internal agronomic factors include variety choice, crop spacing, land availability and rotational requirements.

External factors: are those that the grower has little control over. For example, the prevailing market rather than the grower sets prices. The general political and economic climate will also affect prices and sterling's fluctuation against the Euro, resulting in relatively higher or lower subsidy levels than those in the Euro zone. Other external factors, such as weather conditions, will also have a major direct and practical bearing on the farm business.

Relationships between enterprises: In organic-farming systems, it is especially important to consider the economics of whole rotations and the interactions between the crop enterprises. One way of looking at this is to consider the relationships between different activities or enterprises within a rotation. These can be seen as *competitive*, *complementary* or *supplementary*.

- *Competitive* enterprises compete for the same resources. This is the case for many vegetable enterprises, which might be competing for labour or machinery at the same time, or two crops in a rotation, which compete for the same nutrients.
- *Complementary* enterprises assist each other. For example, a break crop, such as peas, might rest the land, improve the structure and fertility of

the soil and enable the land to be cleaned of a particular disease or weed, to the advantage of a subsequent crop (say, potatoes or brassicas). They may also supply a by-product for livestock (which in turn provides fertility for cash-crop production). In the case of fertility-building crops, this phase of the rotation can be considered a part of the 'costs' of achieving high returns for, perhaps, potatoes or carrots later in the rotation.
- *Supplementary* enterprises imply that the increased production of one crop or enterprise has no effect at all on the production of another. This indicates the use of spare resources within a rotation. A particular enterprise may, for example, use permanent labour in an otherwise slack period.

Looking at the various activities, complementary relationships between enterprises would seem to lie at the heart of the *holistic and systems approach* to agriculture, to which organic farming aspires. Competitive activities are, to some extent, unavoidable from a biological/ecological point of view. It is these types of interactions that, above all, require an economic analysis over the whole rotation period. This view highlights how it can be inappropriate to consider the economics of a single enterprise, especially a high-value crop, such as organic carrots, outside the context of the whole-farm system.

Individual enterprise performances can be positively or negatively correlated and some activity that would optimize gross margins for a particular enterprise would not necessarily optimize the profit or sustainability of the farm as a whole. Thus, it might make sense to not have the best possible gross margins for an individual enterprise, because of detrimental effects in other enterprises.

ECONOMIC ANALYSIS METHODS

In order to assess the technical and economic efficiency of conventional farm businesses, a variety of management techniques or tools have been developed. These include whole-farm account, gross and net margin, and full-cost accounting analyses. All these techniques are also appropriate in an organic farming business, although they all have their limitations. Organic systems require the integration of a number of enterprises and gross and net margins for particular enterprises taken out of the whole farm or rotational context can, therefore, be misleading. Consequently, it is important that any detailed economic scrutiny of an organic system includes whole-farm economic analysis.

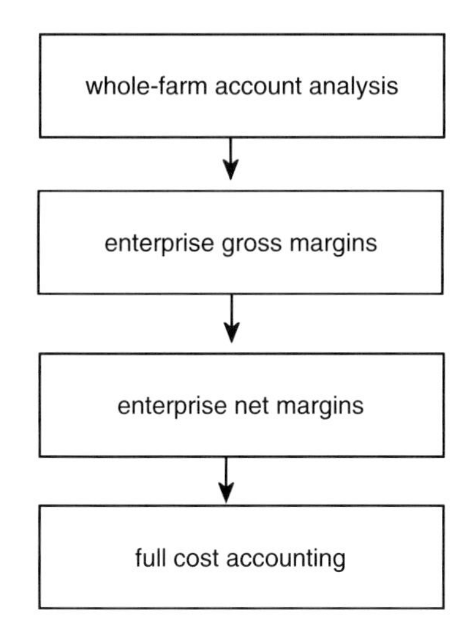

Figure 11.2 Farm economic analysis

Whole-Farm Economic Analysis

An analysis of the farm's economic performance, usually utilizing the farm accounts prepared for taxation purposes, often by an accountant, is a good starting point. These accounts will comprise a profit and loss or trading account and a balance sheet for the year (*see* Table 11.1).

The profit and loss account comprises the sales and expenses for the year and the balance sheet the assets (land, machinery, livestock) and liabilities (money owed) of the business on the date of the year end. Taking the accounts we can observe whether the business has made a profit or loss in the year, by subtracting costs from income. The accounts will also show some breakdown or detail of income and costs, so that we can observe where most of these are being incurred. It is also possible to see how each of these has compared with the previous year and then, if necessary and possible, try to find ways of boosting income or reducing costs.

Benefits: This information is already readily available on the farm. The results can, with some adjustments, be compared with those for other farms. This is often called benchmarking. Farm accounts for the whole of the UK and each region and farm type are collected under the Farm Business Survey (FBS). Specialized studies are undertaken into organic farms. Adjustment to farms' accounts required for comparative purposes are:

- all farms are treated as tenanted, so a rent figure will be added if the farm is owned

Table 11.1 Example of a typical profit and loss account (year ending 30 Mar 2005)

	2005	2004
Sales	£	£
Cattle	6,500	7,500
Vegetables	64,000	48,000
Grants and subsidies	2,000	1,800
Sundry income	750	500
Total sales	**73,250**	**57,800**
Costs		
Wages	6,005	4,867
Livestock costs	800	2,000
Seeds and fertilizers	9,000	6,500
Machinery expenses	3,000	4,300
Rent	5,400	5,400
Water	700	695
Heat, power and light	981	1,500
Insurance	2,000	1,600
Office costs	1,300	1,100
Professional fees	630	600
Bank and loan interest	1,200	1,400
Depreciation	10,000	10,500
Total expenses	**41,016**	**40,462**
Net profit for year	**32,234**	**17,338**

- all interest received and charged is deducted
- all unpaid labour, including that of the farmer, is costed-in.

Limitations: There is often insufficient detail of income and costs to discover why certain enterprises are not performing well. It can be difficult to trace details on key crops, or the source and reason of high or low variable costs. It is, therefore, only of limited use as a tool for farm business planning.

Enterprise Gross Margins

A gross margin is usually calculated for individual enterprises (say, vegetable crops) by subtracting variable costs from financial output (*see* Table 11.2 for an example). The use of gross margins has been widespread in the UK since about 1960, when it was first popularized amongst farm management advisers for analysis and planning purposes. The gross

Table 11.2 Gross and net margins of organic main-crop potatoes per hectare

Crop	Potatoes
Marketable yield (t)	24
Price (£/t)	248
Total Output (£/ha)	**5,952**
Seeds/Transplants	901
Fertilizers (FYM)	79
Crop protection: spray/pest control	156
Casual labour: hand planting	35
hand weeding	92
hand harvest/grade	887
Packaging, storage	249
Transport	158
Commission, levies	8
Other	162
Total Variable Costs (£/ha)	**2,727**
Gross Margin (£/ha)	**3,225**
Cultivations	189
Planting/drilling	105
Mechanical weeding	105
Fertilizing (FYM)	41
Spray, P&D control	129
Irrigation	114
Harvesting	291
Other	20
Total Allocated Fixed-Costs (£/ha)	**994**
Net Margin (£/ha)	**2,231**

Averages of thirteen potato crops grown at one site and monitored by HDRA between 1998 and 2002

margin per hectare – or per head for crops and livestock – can be compared with 'standards'. These are published averages of what might be typically possible in average conditions and are calculated from data obtained from a number of farms. Standard gross margins for organic crops and livestock were first produced in 1994 and are now available in the Organic Farm Management Handbook (*see* Further Reading), which is updated on a regular basis. With each subsequent edition, the data have been expanded and updated to include additional enterprises and have become more robust.

Benefits: In organic systems, gross margins are most useful for farm planning and for making comparisons of enterprises, either on the same farm, between organic holdings, or between conventional and organic enterprises. Table 11.3 presents typical gross margins for selected horticultural crops, as sold to packers, as an example of the type of information available for comparisons. In this context, it is worth noting that direct marketing could achieve higher gross margins.

Limitations: Gross margins exclude fixed costs (general costs that can not be allocated to particular enterprises). This omission needs to be taken into consideration for a more realistic assessment of the farm. Comparison of gross margins between enterprises with different fixed-cost structures can be misleading, particularly when conventional variable costs (say, weed control by herbicides) have been substituted by allocated fixed-costs (such as machinery and labour for weeding) in the organic context.

Table 11.3 Typical gross margins for selected horticultural crops

Crop	Marketable yield (t/ha)	Packer prices (£/t)	Total variable costs for packer market (£/ha)	Gross margin for packer market (£/ha)
Main potatoes	25	220	2,602	2,898
Early potatoes	12	400	2,131	2,669
Carrots	36	300	3,888	6,912
Parsnips	18	450	1,978	6,122
Beetroots	25	200	2,871	2,129
Table swedes	25	350	1,838	6,912
Leeks	12	800	8,015	1,585
Onions	20	275	4,051	1,449
Cauliflower(1)	1,250	5	3,963	2,287
Calabrese	5	900	2,882	1,618
Lettuce – Little Gem(1)	6,500	2.4	7,938	7,662
Lettuce – Iceberg(1)	3,000	3	6,085	2,915
Courgettes	7.5	800	1,655	4,345
Celery(1)	3,750	5	8,690	10,060
Sweetcorn(2)	30,000	0.2	2,360	3,640
Savoy cabbage(3)	30,000	0.45	6,601	6,899
Summer cabbage	25	270	5,356	1,394
Red cabbage	25	350	5,183	3,567
White cabbage	35	300	6,045	4,455

(1) dozens per ha : £ per dozen, (2) cobs per ha : £ per cob, (3) heads per ha : £ per head
2004, HDRA

Certain inputs applied on a rotational basis and with residual effects on subsequent crops, such as organic manures, lime and compost, should have their costs spread over the rotation. It is unrealistic to expect their costs to be carried by the individual enterprise to which they are first applied. Similarly, high individual-enterprise gross margins from cash crops, such as vegetables, do not reflect the potentially very different enterprise mix over the rotation and the need for fertility-building crops in the rotation.

Gross margins should only be compared with figures from farms that have similar production systems (that is, similar soil, crop and/or livestock types). With this reservation in mind, such comparisons can give a useful idea of the production and economic efficiency of any particular enterprise.

Enterprise Net Margins

A further development to the gross-margin method is the subtraction of easily allocatable fixed costs, such as those related to field operations in crop production, to obtain a net margin per crop or enterprise (*see* Table 11.2).

Benefits: This technique has been adopted in a number of studies investigating the economics of organic farming and it overcomes some of the limitations of gross margins (especially changes in fixed costs when comparing organic with conventional farming).

Limitations: The difficulties of using net margins are many. Firstly, there are few (if any) published 'standards' with which to compare a particular farm's net margins. Secondly, costs of field operations are not accurately recorded on all farms (in terms of labour time and fuel usage of machinery), therefore they rely on estimates that may vary from farm to farm and which can cause problems when farm comparisons are made. This can be overcome by using contractor's charges to arrive at the costs of various operations.

Full-Cost Approach

In full-cost accounting or complete-enterprise costing, not only are the outputs and variable costs allocated to individual enterprises, as for gross margins, but all the fixed costs are also allocated. This results in a net profit per enterprise.

Benefits: As all the costs are allocated, it enables the calculation of costs per unit of output (for instance, tonnes of carrots or heads of cabbage) produced on the farm, as well as the calculation of break-even budgets. The strength of such techniques is that they help to identify all the costs involved in a particular enterprise. Such costings are normally required where contract prices are being fixed.

Limitations: Despite its apparent simplicity, the full-cost approach is fraught with difficulties. Awkward, and often arbitrary, decisions have to be made concerning the allocation of overhead expenses between enterprises. For simplicity, this is often done on a per hectare basis but, for accuracy, labour and machinery costs need to be recorded in great detail and then carefully allocated. All this is likely to result in additional paperwork and time. Net-profit figures per enterprise tend to ignore the interrelated nature of enterprises and are, thus, less useful for organic systems. They are of most use where a farm has one core enterprise, such as top fruit or market gardening, and the other enterprises are either minor or seen as contributing to the core enterprise.

Rotational Gross and Net Margin Approach

An assessment of rotations can be performed using gross margins and can be used to compare the profitability of different rotations. Table 11.4 illustrates such a procedure for three typical vegetable rotations. The average gross margins over the rotation are lower than the gross margins of the vegetable cash crops because the fertility-building crops show a negative gross margin and the break crop (spring barley in this case) had a much reduced gross margin. However, the benefits arising from the rotation (fertility, ley establishment under barley, weed, pest and disease control) can make it an indispensable practice. The same analysis can be made using crop net margins, if available.

Table 11.4 also highlights the importance of including fertility building into assessments of the whole rotation performance, despite the fact that its impact on financial performance is likely to be negative (as at least 20 per cent of land is out of cash-crop production at any one time). This is, however, probably unavoidable, because brassicas, potatoes and other vegetables will need to be provided with these intervals for agronomic reasons (*see* Chapter 7). The cost of fertility building is £100/ha (in this example), whereas a failed or un-marketable vegetable crop can cost up to £2,000/ha (seeds, casual labour, and so on). Fertility building can, therefore, also be seen as a risk-minimizing strategy that should be treated as an integral part of organic farming.

Benefits: Enables a full-picture view over the whole rotation, including the fertility-building phase and, hence, helps planning and optimizing rotations in the most appropriate way for vegetable growers. Rotational gross margins can also account for costs not allocatable to single crops, such as sub-soiling, manure applications or lime spreading.

Table 11.4 Crop and rotational gross margins in three rotations

Year	Rotation A	£/ha	Rotation B	£/ha	Rotation C	£/ha
1	Grass/ Clover	–100	Grass/ Clover	–100	Grass/ Clover	–100
2	Grass/ Clover	–100	Summer cabbage	1,394	Grass/ Clover	–100
3	Potatoes	2,898	Onions	1,449	Lettuce (double crop)	10,576
4	Carrots	6,912	Barley	478	Leeks	1,585
5	Barley	478				
	Average (£/ha/year)	**2,017**		**793**		**2,990**

Limitations: Similar to gross and net margins, it excludes fixed costs. It also excludes non-rotational enterprises, such as livestock, and therefore limits the use for a whole-farm economic analysis in certain farm types.

ECONOMICS OF ORGANIC VEGETABLE GROWING

This section provides a detailed look into some of the important aspects of the economics of organic vegetable growing, such as the conversion and post-conversion period, rotational planning, risk assessment and benchmarking.

Vegetable systems are generally more intensive than other agricultural systems. This means that they require greater inputs per land unit to sustain the greater output, potentially leading to a higher income from a small area. Returns, however, are notoriously variable, making it important to carefully analyse the farm business. These systems are riskier because the product in question – fresh vegetables – is perishable and, thus, presents a very limited time-span in which it is marketable. In addition, the high costs involved in vegetable production imply that there is potential for large losses if the produce is not marketable.

Compared to conventional production, inputs to organic vegetable-production are generally higher, apart from the reduction in costs associated with pesticide and fertilizer application. For example, the cost of most organic vegetable seeds and transplants is significantly higher than their conventional counterparts. These usually make up for 10–40 per cent of variable costs. Higher input costs are also found for casual labour, especially for hand weeding. Increased numbers of passes needed with mechanical weeders result in higher allocated fixed costs. However, weed-control costs can vary greatly depending on the crop, the weather, equipment available and skill of the grower. Indeed, labour is a major area of input costs in organic vegetable production. In common with conventional production, organic farms often make use of gang or contracted labour. There can be difficulty in sourcing labour in some parts of the country. In response to this, there is evidence that organic farmers are developing improved mechanical-weeding systems in order to reduce casual labour, but such shortages in good quality casual labour could be a constraint to the future expansion of organic vegetable production in the UK.

Conversion

Economic sustainability is particularly sensitive during the conversion period (*see* Chapter 2), since land will generally have to be taken out of effective production for at least two seasons, in order to build soil fertility and satisfy the regulatory authorities. This change in system orientation involves a large investment (money, time and foregone sales) that will only pay off in subsequent years, when the land is fully organic. This, in turn, can lead to a substantial reduction in cash flow during this period, even if crops are grown, as there is a very limited market for 'in-conversion' vegetables. The key to a successful conversion is to build fertility, and to phase it in over a number of years, allowing experience to be gained gradually. However, costs can also be minimized with careful planning and examination of costs prior to conversion. Farm finances can be aided by taking advantage of any financial aid that is available from governmental programmes.

Research undertaken over a number of years at HDRA has shown that the degree of decline in income during the two-year conversion on a farm is related to the level of intensity of production prior to conversion and to the proportion of land entering conversion each year, or the rate of conversion. This reduction in income during conversion is combined with the need to invest in new equipment for tasks such as mechanical weeding. On one 200ha mixed-cropping farm, converting to organic vegetable production was estimated to cost around £150/ha/year. The total 'costs of conversion', including the reduction in income on the same farm, were estimated to be £550/ha. The overall costs of conversion can in many cases be offset by claming organic aid (for example £450/ha in 2004–05 over five years in England, subject to change in the future) and set-aside payments on the grass/clover leys. Where land is not registered for arable area payments, for example on more intensive vegetable farms, set-aside payments are not available and organic aid is payable at a lower rate, thus pushing up the 'costs of conversion'. It should be borne in mind that there is considerable variation between farms, with whole-farm conversions and those converting intensive vegetable systems having the greatest costs.

Post-Conversion

The economics of organic vegetable systems following conversion are dependent on yields, prices, costs of inputs and the amount of fertility-building crops in the rotation. Immediately following conversion, yields of vegetables are likely to be 20–50 per cent below conventional levels, although in subsequent years yield levels generally increase. This increase is partly due to greater proficiency in organic production methods as growers learn and adapt their production system, but also due to enhanced soil-fertility as the organic system becomes more established.

The economics of organic production are very sensitive to prices. For example, in 1999 and 2000, prices for organic vegetables were commonly double those of conventional ones, enabling farmers to obtain financial outputs comparable to those achieved in conventional production. Research has shown that on monitored farms, in this period, the gross margins achieved were equivalent to, or in some cases higher than, comparable conventional levels. However, since 2001, the price of organic vegetables has fallen on average by 20–25 per cent, as supply has increased, which has led to a reduction in profit. At the same time, marketing has become a bigger challenge for organic growers, as opposed to the technical constraints to production (*see* Chapter 10).

An additional input to organic vegetable production is the need to have a proportion of the land under fertility-building crops. In stockless vegetable-growing systems, restoring fertility, in the absence of chemical fertilizers, is achieved by growing legume crops, commonly grass/clover leys and/or over-wintered green manures. These fertility-building crops may occupy 25–50 per cent of a typical organic vegetable rotation. In terms of economics, this cost needs to be considered over the whole rotation, whereby the average gross margin for the rotation incorporates the cost of the fertility building (*see* Table 11.4). Research has also shown that costs for fertility building are estimated – as a 'rule of thumb' – at about £1 per kg N/ha supplied to the rotation. For example, short-term fertility building (six to twelve months) may supply 100kg/ha and cost about £100/ha and two-year grass/clover leys may produce 200kg/ha and cost about £200/ha to establish and maintain. In both cases, organic seed costs are the main costs. There is scope to optimize rotational planning and details on a step-by-step approach on a farm level are given in the following section.

Rotational Planning

Rotational planning is one of the major management issues for a successful organic vegetable-growing enterprise. Growing the wrong crops with no market, insufficient long-term fertility building and build up of large weed seed banks can all be reasons for failure of an organic enterprise. Financial analysis, as outlined, is one tool that can help with decision-making. It can help to 'measure' or, better, attach a value to all farm operations, either on a crop level as 'crop gross margin', or on a rotational level as 'rotational gross margin'. On an environmental level, a cost-benefit analysis can compare the farm-level economics against the environmental costs and benefits. In planning a rotation, financial information, either collected over previous seasons or estimated from standard sources, can help make decisions about rotation design. This emphasizes the importance of keeping and analysing financial information, apart from the necessity to keep such information for legal business accountancy and tax purposes.

Planning for the rotation will need to take into account the following different factors:

- market (demand, price and specification)
- skills and expertise available, as well as machinery and resources
- nitrogen management
- management of other nutrients (e.g. P, K)
- weed management
- soil fertility-building (structure, soil fauna)
- pest and diseases issues
- irrigation
- environmental issues (N and P leaching, agro-ecology, energy use).

A series of steps is suggested below for planning and modifying a rotation, incorporating an element of economic analysis:

1. Gather knowledge of all the cash crops that it is possible to grow on a given soil type. This should include all crops, even those for which there is no cropping experience. Experience and expert knowledge can serve as a useful guide at this stage.
2. Gather knowledge of all possible fertility-building crops. This should include long-term and short-term fertility-building crops (leguminous and non-leguminous). Expert knowledge as well as experience will need to be used in coming to a choice.
3. Depending on market conditions and current cropping experience, select key (cash) vegetable crops to grow. Once this decision is made, the other factors and crops are grouped around this decision.

4. Assess any technical necessities and constraints, like irrigation or accessibility of harvest equipment. Adjust the list of cash crops depending on the constraints identified. Financial constraints might also exist, for example limits to possible investment in appropriate machinery.
5. Choose suitable fertility-building crops. The most suitable will depend on whether long-term (two or more years), medium-term (one year) or short-term fertility-building periods are used. A combination of all three is often a good strategy, although the details will depend on the amount of fertility-building in the rotation at any one time and the combination of cash crops (for instance, intervals between brassicas). Fertility building can include innovative techniques, like undersowing, wide-rows or companion planting.
6. Calculate (or estimate) the rotational gross margins of different possible combinations using the Organic Farm Management Handbook (*see* Table 11.4). It is important to note that the economics of the whole rotation is taken into account, including fertility-building crops, possible double cropping and/or fallow periods.

It might not always be possible to buy state-of-the-art machinery like this leek dibber, due to financial constraints

7. In the next step, the rotations with the highest rotational gross-margins are assessed further with respect of their possible implications on weeds, pest and diseases, and nutrient use. For this assessment, again, expert knowledge is needed, because interactions of these factors with climate and soil type are variable and difficult to predict.
8. The environmental assessment is the final step. This is not because it is the least important, rather because all other factors need to be balanced first. If significant negative agro-ecological effects are found, some fine tuning of the rotation might be possible, but it is also possible that the whole process needs to be repeated.

Obviously, many of these steps will be interdependent and should also constantly be under review in established organic-rotations. Advice is available from advisory services to help make decisions and should be regarded as a necessary investment cost, especially when planning new enterprises (*see* Chapter 13). Increasingly, computer software is becoming available for decision support on rotational planning, which, theoretically, allows a number of different rotations to be evaluated very rapidly using farm financial data and expert knowledge (built into the model). Two current models include OrgPlan (for which there is a modest licence-charge) and EU-Rotate-N (which is free but under development) (*see* Useful Addresses).

Risk Assessment and Avoidance Strategies

Agriculture and horticulture are, by nature, risky activities. They rely on factors, such as weather, that are beyond the control of producers and, therefore, carry extremely variable outcomes. In addition, variability in yield, price, labour use and marketing costs magnify risks in agriculture. For this reason, the results of economic analyses and benchmarking data are best considered with care.

The strategies applied by growers will vary considerably with their aims, objectives and attitude towards risk. While one grower might be risk-averse and prefer a stable income, another might be more willing to take a risky activity that is more likely to fail but could eventually lead to larger rewards.

Organic growers often pursue strategies, as a means of spreading or minimizing risk, that are suited to the ethos of organic farming and its underlying principles. Some of those strategies are:

- *Diversification* spreads risk, as the more crops there are, the lower the probability of having a total failure. (If one crop fails, there will be others on which the grower can rely.)
- *Fertility building* and rotational pest and weed control

- *Direct marketing* reduces risk associated with the variable, and often rather low, prices offered by supermarkets. (It allows growers to gain more control over the price of their produce.)
- *Hedging* of prices or costs at futures markets (say, potatoes or fuel costs)
- *Contract growing* with fixed prices (*see* Chapter 10).

Whole-Farm Budgeting

One of the greatest benefits of economic analysis is that it provides a business with information, which can be used to construct a whole farm

Table 11.5 Budgeted income and costs on a 13ha (33 acre) vegetable farm

INCOME(1)	**ha**		**£/ha**		**£**
Potatoes	2		5,000		10,000
Carrots	2		10,000		20,000
Mixed vegetables	6		12,000		72,000
Other income(2)					1,000
TOTAL INCOME					**73,000**
Variable/ direct costs	**t/ha**	**£/t**	**£/ha**	**£**	
Seeds and plants			494	4,942	
Composts and fertilizers	25	10	247	2,471	
Packaging			297	2,965	
Horticultural sundries			618	6,178	
Total variable costs			1656	16,556	
GROSS MARGIN					**56,444**
Fixed costs	**£/h**	**hrs/ha**	**£/ha**		
Labour	6	750	4,500	27,000	
Irrigation			247	2,471	
Rent			247	2,471	
Other overheads			1,236	12,355	
Machinery repairs				1,500	
Fuel & oil				1,500	
Machinery depreciation(3)				2,000	
Polytunnel depreciation(4)				1,350	
Interest on borrowed capital				300	
General repairs				500	
Total fixed costs				**51,447**	
NET PROFIT					**4,997**

(1) excluding 25 per cent of land under fertility building, (2) for example, subsidies and grants, (3) 20 per cent of value of all equipment, (4) 30 per cent on costs of these

budget. This is an important planning tool. The budget will be constructed using plans on the area of each crop or other enterprises on the farm. Specific yields, prices and costs can then be fixed to these. Ideally, these should be obtained from the farm's own performance, or if not, from sources such as farm-management handbooks. The budget can be constructed as in Table 11.5 or, alternatively, individual gross margins can be used and assembled, together with fixed costs. Future operations and business development can then be anticipated and planned for, reducing some of the inherent risk in organic vegetable production.

In conclusion, it is important for any agricultural business to have an on-going process of planning, recording, analysing and re-planning. This will involve planning rotation, resource allocations and enterprise mix. Recording and analysing performance will allow an assessment of business 'strengths and weaknesses' so that it can be continually improved.

Chapter 12

Vegetable Crops for Organic Production

Whilst a good grasp of the overreaching principles and practices of organic vegetable and potato production are necessary, it is also important to appreciate that growing any particular crop type successfully is likely to require further detailed and specific knowledge. Indeed, all vegetable-growing systems will have to be tailored, to some extent, to the types of vegetables grown, as well as to the resources, such as skills, materials and finances, available to grow them. In practice, this means accommodating and balancing the various principles and practices outlined in the previous chapters with these resources. The aim of this chapter is, therefore, to synthesize many of the themes developed in previous chapters by concentrating on specific crops (or crop types) to give some insight into crop production methods that are likely to be used for the most commonly grown vegetables. Whilst it is not possible, and generally beyond the scope of this book, to give a highly detailed description of all production techniques for all the various possible temperate organic vegetable crops, here we have aimed to highlight the production practices and problems that growers will be most likely to face in their day-to-day activities and, by extension, give them an idea of what they might need to think about or plan for when taking on new crops.

A brief overview of the crop families is given in Chapter 3 and we have taken this as a starting point for this chapter. In presenting the information, we have divided vegetables into seven broad types: leafy salads, leaf and flower vegetables, fruit vegetables, pod and grain vegetables, bulb or stem vegetables, root vegetables and perennial vegetables. Each vegetable type covers a range of families and species, which are characterized in more detail in Chapter 3 and summarized in Table 3.1. Categorized in this way, the types represent groups of vegetables that usually require similar husbandry and/or marketing techniques and so this classification represents a practical grouping only. In each category some of the more commonly grown or staple crops are described in detail, in order to demonstrate the principal husbandry techniques involved. In addition to this, we have provided an overview of the main points involved in producing some of the less commonly grown crop types to give an idea of the possibilities for developing diverse vegetable-production systems. It should be borne in mind that care should be taken when designing rotations, to observe the required intervals between botanically similar crop families, even though they might fall under the same type-category in this chapter.

Highly detailed information on pests, diseases and weeds is not given in this text but, if it is needed, it is generally readily available in specialist books or leaflets (*see* Useful Addresses and Further Reading). Here we have merely indicated the most likely that will be encountered in any cropping situation (*see also* Chapter 5 for details on weed management and Chapter 6 for details on pest management). Likewise, soil-fertility management has been dealt with at length in Chapter 4 and only specific details are mentioned in this chapter. As an aid to rotational design the botanical family group has been given, along with the Latin name for each crop. The crop family has been coded as: All – Alliums (Amaryllidaceae), Br – Cruciferae (Brassicaceae), Ch – Chenopodiaceae, Comp – Compositae (Asteraceae), Cuc – Cucurbitaceae, Leg – Leguminosae (Fabaceae), Sol – Solanaceae, Um – Umbelliferae (Apiaceae), O – other.

LEAFY SALADS

'Leafy salads' covers a wide range of crops, increasingly in demand as salad packs become more popular. The wide range available means that salad packs can be tailored towards very varied tastes. The types and varieties used come from a number of different vegetable (and herb) families. Some of the more common salad crops are described below, starting with the most familiar types: lettuce, endive, chicory, oriental brassicas, oriental mustards and Pak Choi. These types cover the range of growing techniques that are likely to be needed to produce all salads and other salad crops, like rocket and baby-leaf spinach, are described in less detail at the end of this section. More detailed knowledge on their cultivation will need to be found in more specialist texts.

Many of the leafy salads are only in the ground for short periods of time and can be useful to fill in gaps in a rotation or for extending the season either side (namely, early and late). Some crops are cut at the baby-leaf stage (spinach, lettuce, beetroot, leaf beet) and others are left to mature until the whole head or plant is cut. In some cases, individual leaves can be harvested from plants and the plant left to grow on and produce more. New salad crops are constantly being introduced, especially from Asia, which is why many seed companies list them under oriental vegetables and salads. Many of these species were initially introduced as alternatives to lettuce during the winter months and are, therefore, cold tolerant and quick to bolt in hot weather. They are particularly suitable for growing under protection in the winter, to provide mixed-salad packs in the autumn and late winter.

In addition to salad crops, the young tops of various other crops can also be used in salad packs. These include tops from radish, turnip, kohl rabi, beetroot and onions. In this case, if removing foliage for salad packs, it is necessary to leave enough leaf to allow the remaining crop to feed itself and develop. Herbs are a distinct group of (generally) leaf crops that are used for flavour. Some of these crops require specialist and detailed knowledge for their cultivation, whilst others are relatively easy and trouble free. Some of the leafy herb types have been briefly mentioned in the section on other leafy salad vegetables.

Salad crops represent a wide range of different plant families and care needs to be taken when designing rotations. The brassica family is well represented and organic standards specify that four years should be left between brassica crops in a rotation. The brassicas share an extensive range of pest and disease problems, which might differ in importance between the crop types, depending on which parts of the plant are marketed, although oriental brassicas tend to escape them altogether.

Lettuce (Comp, *Lactuca sativa*)

Lettuce is the most commonly available salad leaf and especially popular in the summer months. Demand can, however, be highly variable according to the season. Generally cut and sold as a whole head, it is also increasingly being cut as leaves (for salad packs). All types of lettuce can also be grown as a cut-and-come-again crop for baby-leaf production. For salad packs, the leaves of bolted lettuce are still usable, although for a limited period as they soon become bitter and tough. With protection and some heat in the winter, it is possible to crop all year round, using appropriate varieties, although timing is crucial due to the varying speeds at which heads mature at different times of year. Without heat it is still possible to crop for most of the year, as a protected winter-crop will stand well. Lettuce for salad leaves can be available all year round, even if the plants do not form heads.

Types and Varieties: The most important characteristics for organic growing are probably vigour and disease resistance. Varietal resistance to various diseases – especially mildew (*Bremia lactucae*) and viruses – is available and some varieties are resistant to some pests. A range of lettuce types is available; butterhead, Batavian, crisp (including Iceberg), looseleaf and Romaine/Cos (including Little Gems). All differ to some extent, as regards leaf- and head-size, shape and texture, and some characteristics of each are given below.

Batavia: these lettuces are a cross between crisp and butterhead types (*see* below), with some very useful new varieties showing lots of vigour and standing ability and achieving good head-weights. They can be green or red, looseleaf or heading and are very popular with customers.
Butterhead: these are useful for early and protected cropping because they have a relatively short time to maturity. However, their popularity has declined in recent years. Their softer leaves wilt sooner than other varieties, meaning a shorter shelf-life, and they have less potential for trimming off the damaged or diseased outer leaves. They are also quicker to bolt than other types.
Crisp/Iceberg: Icebergs are possibly the most popular lettuce for the supermarket trade, although they are of less use for salad packs, as the leaves cannot be separated easily. They can be trimmed hard, which is an advantage where disease symptoms and slug damage are evident, but they have the longest growth period and require good irrigation to make the head weight. Choose varieties specifically bred for Iceberg production, as opposed to the Webb's Wonderful types that are sold with the large outer leaves left on.
Little Gem: this relative of the Romaine types has softer leaves and is much smaller in size. Previously very popular as a pack of two or three heads on supermarket shelves, their popularity has declined in recent years. They are good for salad packs, due to their small leaves. Spacing between plants is less than for other lettuce types but, otherwise, cultivation is the same.
Looseleaf: these include Lollos, Salad Bowls and Oakleaf types. There are red and green varieties of each type available. They are very popular for inclusion in salad packs and, to that end, are easy to grow, as it is not necessary for them to reach a set head-weight. They are not as easy to market as whole heads, however, except perhaps to the restaurant trade as part of a mixed box.
Romaine/Cos: lettuces are very upright and tough with a sweet flavour. They are not very well suited to salad packs, as leaves are too large, although heads that don't achieve a good weight can be used. They are well suited to organic cultivation and are often more tolerant of cold conditions.

Seed and Raising: There is a wide range of organic seed available and varieties are constantly being improved, with new varieties appearing frequently. Lettuce seed is small and very hard to handle; therefore, for large quantities, it probably pays to buy pelleted seed, especially if sowing

Pinokkio, a type of cos lettuce, ready for harvest

by hand. The only drawback to pelleted seed is the difficulty seeds can have breaking through the coating if it dries out.

Transplant raising: Lettuce can be direct-drilled into a fine nursery-seedbed and transplanted at the three to five true leaf stage. However, the most common commercial method is to raise transplants in blocks or modules. Raising transplants is essential for early crops, before soil temperatures rise enough for germination – the optimum temperature being 13–16°C. In the summer it may be necessary to cover trays to prevent over-heating, until germination. Using transplants also reduces the risk of seedling loss to slugs and other predators during their most vulnerable stages. The optimum module volume is about 25cl, alternatively they can be raised in blocks of 2.5–4.5sq cm (0.4–0.7sq in). The transplants will not hold in the trays for more than a few days and there may be problems with mildew at high plant-densities. Trays must be kept well ventilated and not too wet, especially overnight, and the plants must be kept moving, with supplementary feeding if necessary. If damping off or mildew is a problem, then hygiene needs to be stepped up, trays sterilized (with an approved solution) and, if possible, the batches kept separate.

Direct drilling: this is only common for baby-leaf crops (which can be turned around in sixty days at peak season), although heading lettuce can be drilled at 70–100mm (3–4in) and thinned to 300–350mm (12–14in). Pelleted seed makes the task hugely more manageable.

Growing: The bed system is the most common method of growing lettuce. It reduces compaction and a fine tilth is necessary to ensure good contact with the block or module and to facilitate weed control. Standard planting machinery is available. It is important not to plant too deeply (burying the growing point), which leads to misshapen heads and risks soil contaminating the heart during weeding operations, which can, in turn, cause rots. Planting too shallowly leads to drying out.

Spacing varies according to variety, with Little Gem types at 150mm (6in), Cos, butterhead and looseleaf at 300mm (12in) and crisp varieties at 350–440mm (14–17in) in the row, with about 300mm (12in) between rows. Other considerations, in terms of spacing, are head-size requirement (wider spacing increases size, narrower spacing decreases it), fertility levels and weed versus disease issues (closed canopy for suppression of weeds versus open canopy to increase ventilation and suppress disease). Access for weed control needs to be considered too.

As a rule of thumb, successional sowings need to be made at roughly two-week intervals (slightly longer at the beginning of the season and shorter in mid-summer) and undercover crops need to be in the ground by mid-September if they are to have a chance to be mature when the outdoor crops finish in early December. Outdoor cropping should be possible from late May with the aid of fleece (although this should be removed several weeks before harvest to prevent leaf damage) and will stand until a fairly heavy frost. Lettuce does not tolerate heat well, however, so will not stand long in hot weather before bolting.

Protected Cropping: Although the breeding of short-day varieties (which should always be chosen for winter cropping) has greatly improved over-wintered protected crops, they require much longer growing periods as compared to outdoor lettuce (unless heat is available). To have heads ready before Christmas, they need to be planted by the end of September, with the same fertility levels as for outdoor crops. On very sunny days and/or as spring advances, good ventilation is essential during the middle of the day, to prevent pest and disease build-up. The same applies for all year round undercover production, although for the summer period long-day varieties should be used. As well as the pest and disease problems encountered on outdoor crops, springtails may be a problem and slug numbers are likely to be high. The high-value nature of the crop makes it cost effective to introduce a biological control (the nematode *Phasmarhabditis hermaphrodita)* to combat slugs. However, soil temperatures must not be allowed to drop below 5°C. Plant spacing can be reduced slightly to 220–250mm (8.5–10in) for most varieties.

Soil Fertility and Rotation: There are no limits set by organic standards for rotational period between crops but, in terms of pest and disease management, at least a two-year gap between crops is good practice. Where ringspot (*Microdochium panattonianum*) or *Sclerotinia* rot have occurred, a five-year gap is preferable.

Lettuce requires medium soil fertility, so is best grown either after a ley or after an application of FYM or other fertility source. It doesn't perform as well on dry soils or on heavy clays, preferring good structure and drainage. A pH of 6.2–7.0 is optimal, as is a high organic matter content to increase water retention.

Irrigation: Lettuce is a thirsty crop, therefore it is best not to plant in a dry situation and to be vigilant with irrigation during dry weather. The best forms are trickle or drip irrigation, to reduce wetting of the leaves.

Weeds: Once the crop is developing, it is very difficult to perform any weeding operations, because of damage to leaves and the risk of contaminating hearts with soil, so weeds need to be under control from the early crop stage. On the plus side, the outer leaves soon spread, resulting in good weed-suppression. A clean seedbed is always beneficial and, after planting, a steerage hoe is probably the best mechanical means of weed control, until the point when the heads are big enough for uprooting to be a risk. Tractor speeds need to be kept down, to avoid throwing soil onto the lettuce. Brush hoes can be useful in wetter conditions (otherwise dust is a problem), though careful setting is necessary to avoid tearing the leaves. Row widths should be kept to 250mm (10in) minimum. Fast-growing crops probably only require one pass with hoes at about three to four weeks after planting or at ten to fourteen days for later crops. Slower crops, for example crisp types, may require a second pass in the row, which would need to be carried out by hand.

On a smaller scale, hand hoeing is very effective and can, with care, be performed at a later growth-stage. Weed control can also be achieved by planting through holes in plastic mulch, although this may encourage slugs and there is a risk of roots overheating in very hot weather.

Pests: The main pests are slugs, cutworms, leaf aphids, root aphids and vertebrate pests, like rabbits and birds. Wireworms and leather jackets can also cause problems, especially after fertility-building leys. Rabbits can be kept off with netting over the crop itself or by fencing the area. Electric fencing needs to be very well maintained and of a high charge (5,000 volts), as rabbits are remarkably tolerant of electric shocks. Permanent wire-mesh fencing is more effective. There is little that can be done to deter slugs, so heads need checking for inhabitants before delivery if there is a significant problem, as they are very off-putting for customers.

Cutworms (larvae of moths, for instance *Agrotis* spp.) can sometimes be found under recently severed heads and destroyed. Otherwise keep crops at risk well-irrigated, as the young larvae and moths die in wet soil and it keeps the crops growing through. There may be a warning system operating in the local area.

Aphids can be a serious problem. Foliar aphids, such as currant lettuce aphid (*Nasonovia ribisnigri*), peach potato aphid (*Myzus persicae*) and potato aphid (*Macrosiphum euphorbiae*), cause some damage to the leaves and can also transmit viruses. However, the greatest damage is usually rejection for marketing purposes, due to a general consumer-intolerance of any aphids in heads. Varieties offering some resistance are available, which aphids seem not to prefer. Soft soap can be sprayed if there is an outbreak, although it requires contact with the pest to be effective and is unlikely to reach any aphids right inside the heads. Isolating successive crops will help prevent the problem from spreading. Maintaining high levels of natural predators in and around the crop can help, as can keeping crops weed-free and destroying debris between successive sowings. Fleece might help, but can damage the delicate heads or cause abnormal growth if left on too long. In successional plantings, it is often common to lose one planting to aphids when the predator/pest balance is out of kilter but, given patience and an understanding market, the predators will usually win out in the end.

Root aphids (*Pemphigus bursarius*) can be found on the roots of lettuce plants in early summer after the aphids have migrated from poplars. The aphid can also persist over winter in soil. Some varieties of lettuce appear to be at least partially resistant to this aphid, which can cause stunted and wilted plants, especially in dry weather. Adequate watering during dry weather, crop rotation and thorough cultivation of crop debris can all help to manage this pest.

Diseases: downy mildew (*Bremia lactucae*) is the most common and problematic disease of lettuce, especially in late crops. It can be controlled through good hygiene, crop separation, spacing for adequate ventilation, resistant varieties (although resistance does break down over time) and good management to keep the crop moving. Foliar treatments (for example, seaweed sprays) to enhance the general health and, thus, resistance of the crop may help, but research results on this are ambiguous and a derogation may be required.

Ring spot (*Michrodochium panattoniana*) is commonly observed in cold, wet conditions, although it is generally prevalent at low levels. It is best managed through good hygiene (as it is a soil-borne disease and present in crop debris), as well as good drainage and at least a two-year break between crops.

Rots or bottom rots (*Rhizoctonia solani*, *Sclerotinia* spp., *Pseudomonas* spp. and *Erwinia carotovora*) are generally caused by soil-borne pathogens, so good husbandry, good hygiene and good ventilation all work as preventatives and, where a problem, longer rotational periods should be observed between crops. Grey mould (*Botrytis cinerea*) is a common rot that usually results from damage to the lettuce, say during weeding operations, so care should be taken when managing the crop.

Viruses are generally seed-borne (such as lettuce mosaic virus (LMV)) or transmitted by aphid vectors (such as cucumber mosaic virus (CMV)) and can also be present in crop weeds such as chickweed. They result in sickly-looking or mottled leaves. Good resistant varieties are available for the more serious LMV. Good weed control, combined with aphid control, can reduce risk of cross virus transmission and contamination.

A final important condition is tipburn, which is related to calcium deficiency and hot, dry conditions. Once again, good husbandry is vital, for example avoiding compacted soils and other stresses that exacerbate the condition. There are also differences in varietal susceptibility. Sudden bursts of plant growth can also bring on the condition. Tipburn frequently leads to secondary rots.

Harvesting: Heads mature, on average, over seven to fourteen days, allowing for several cuts from each sowing. In hot weather, however, they can become unsellable within three to four days. Over the season you can expect to get 60–70 per cent to market, although this varies quite widely with conditions and the target market. Care needs to be taken not to bruise leaves or contaminate heads with soils, both of which can lead to in-store rots. If bagging by hand, the quickest method is to tie the bags on their wicket securely round the waist, so that the head can be inserted into the bag while it is still attached to the wicket. The condition of the cut heads deteriorates rapidly in hot weather, when harvesting should be performed only in the early morning. Aim to remove the heads from the field within half an hour.

Storage: Lettuce leaves can retain considerable field heat after harvest, which means heads require cooling immediately to bring temperatures down to about 3°C. At very low temperatures (about 1°C) heads will hold for five to seven days, although they may require re-trimming before delivery and their shelf life will be reduced. Ideally, however, they will be shipped on the day of harvest.

Chicory (Comp, *Chichorium intybus*)

Chicory leaves are useful for salads and they perform well undercover in winter as a cut-and-come-again leaf and the tight headed cultivars achieve some self-blanching, so are not as bitter as the loose-leaved types. For outdoor crops the bitterness may decline after exposure to frost.

Types and Varieties: There are several types of chicory cultivated for their roots or shoots. The best known of the root types is Witloof chicory, with its forced head or chicon, which is harvested at around 150mm (6in) tall and requires blanching during its main-harvest period of late winter onwards. During the summer, unblanched leaves make a good addition to salads. However, raising Witloof chicory is a specialized and time-consuming activity and is not worth the effort for salad leaves alone (although chicons that don't make the grade can certainly be used). Other root cultivars can be sliced raw into salads.

The best known of the leaf types are the 'Sugarloaf' types, which look a bit like a large Romaine-type lettuce, with a very firm heart and whose heads can achieve so great a stature that it can be hard to fit the leaves into salad bags, and red chicory or radicchio, which forms a short, dense head of red leaves that intensify their colour with the cold. They can be blanched to a pink colour and can be cut as a cut-and-come-again crop or left to mature. There is also the rosette-shaped and very hardy 'Grumolo Verde' and various wild chicories that resemble dandelions.

Sowing, Planting and Growing: heading varieties are best raised as transplants and spaced at 250–300mm (10–12in) in the row, although the loose-leaf cultivars can be direct-drilled or broadcast. All types are traditionally sown in early summer, or late summer for protected cropping, although new varieties can be sown in spring for summer use so that, with a combination of varieties, leaves can be harvested almost all year round. The heading varieties are more cold-tolerant and the asparagus and wild chicories more able to stand the heat of summer.

Soil Fertility and Rotations: the plants are unfussy, adapting to a wide range of soil types, as well as wet, dry and exposed positions. They will tolerate light shade in summer.

Irrigation: chicory is deep rooting with good drought resistance.

Weeds: similar to lettuce (*see* above).

Pests: chicory is unlikely to be seriously affected by any pests, although those that are also found on lettuce may attack it.

Diseases: lettuce ringspot (*Macrodochium panattonianum*) – a fungus – can attack outdoor crops in cold, wet weather. Sclerotinina disease (*Sclerotinia sclerotium* or *S. minor*) can also attack the base of the plants and is most often observed under cool, damp conditions. This disease is best managed by crop rotation, leaving at least three years between susceptible crops and taking care to control weeds from the same family that can also succumb to the disease.

Harvest: Sugarloaf and radicchio can be harvested as cut-and-come-again leaves at 50–100mm (2–4in) or left to mature, grumolos being cut slightly taller at 50–300mm (2–12in) and asparagus types at 200–400mm (8–16in).

Endive (Comp, *Chichorium endivia*)

There are two distinct types of endive, curled and plain-leaved (or Batavian), which is hardier and can tolerate slight frosts, making them ideal for winter cultivation; a period in which they can be a good substitute for lettuce. Leaves have a tendency towards bitterness, so that blanching is a common practice once the plants reach maturity at about twelve weeks after sowing. This can be done by tying leaves together in a bundle or by covering as much of the plant as possible, for instance with an upturned flowerpot with the holes blacked out. Very frizzy types remain mild, however, and do not require blanching and more modern varieties have been bred for self-blanching or sweetness.

Types and Varieties: endive is easily grown and responds well to cut-and-come-again treatment. Summer varieties can be sown successionally from April to August, with frost-tolerant crops sown from August to mid-September to over-winter.

Sowing, Planting and Growing: for direct-drilling or raising transplants and subsequent cultivation, treat as lettuce.

Soil Fertility and Rotation: it prefers a good loamy soil with plenty of organic matter, but otherwise as lettuce.

Irrigation: as lettuce.

Weeds: as lettuce.

Pests: endive is not commonly susceptible to any pests, although those commonly found on lettuce may also be attracted to endive. It may be that the bitterness of the leaves acts as a deterrent.

Diseases: the most common disease of endive is lettuce ringspot (*Macrodochium panattonianum*), which is a fungus that particularly attacks outdoor crops in cold, wet weather. Good drainage and hygiene with a two-year break between crops helps to prevent its occurrence. In the south-west centaurea rust (*Puccinia cyani*) has been observed and may be a problem.

Harvesting and Storage: as endive is often grown as a late-season or over-wintering crop, many of the harvest and storage problems of leafy crops in hot weather are avoided. Generally it should be treated as lettuce.

Chinese Cabbage or Chinese Leaves (Br, *Brassica rapa* var. *pekinensis*)

Chinese leaves are becoming increasingly popular as an alternative to lettuce, especially over the cooler months or winter. They do not generally like too much heat and develop best at between 13–20°C during their main period of growth. They should, therefore, not be sown too early or they will bolt. Undercover they can be used as cut-and-come-again crops or semi-mature heads and, in this case, they will stand temperatures down to –10°C. Bolting is also caused by exposure to low temperatures in the early stages of growth so, in cold periods, seed must be germinated and grown on with heat. It is also necessary to choose a variety appropriate to the season. They will usually tolerate a light frost. Cultivars that tolerate lower temperatures are constantly being sought, with new ones being introduced regularly.

Types and Varieties: Chinese cabbage can be divided into headed (barrel-type or cylindrical) and loose-headed (semi-headed or fluffy-top) cultivars. They also divide between those that are easier to grow and those that are harder. There is a great deal of variation within the groups, with a wide range of types. Headed types have a broad, white midrib and crinkly texture, with pronounced white veins. Their flavour is mild and they are crisp and juicy.

Loose-headed Chinese cabbages are less common than headed varieties, although the flavour is the same. They tend to mature more rapidly, often in about eight weeks (55–70 days). Some are tolerant of higher and lower temperatures than headed varieties and are, therefore, more resistant to bolting. Fluffy-top varieties are less cold-tolerant and early sowings often bolt, therefore for mature heads they should not be sown before early summer. They perform best as semi-mature or cut-and-come-again crops, sown from mid-spring through the summer, with earlier and later sowings undercover.

Sowing, Planting and Growing: the seed leaves of the headed types are often hairy and may not be suited to a baby-leaf crop, although seedling crops are possible with the fastest-maturing varieties and the smoothest leaves. These can be useful where slugs are a problem, as damage is less obvious and it is possible to get up to four cuts, with the first as early as fifteen days after sowing. The first sowings can be made undercover in early spring, as long as soil temperatures are above 5°C and continue so for several weeks, although plants will bolt once day temperatures rise. Outdoor seedling crops can be drilled from late spring to early summer, with undercover sowings recommencing in late autumn or early winter. These will probably only provide one cut before growth ceases, but may survive to re-sprout in late winter.

For a mature crop, headed cultivars should not be sown before early summer and preferably between mid-June and mid-August. Use bolt-resistant barrel types for the earliest sowings. They are best raised as transplants and planted out at the 4–5 true-leaf stage at 300–450mm (12–18in). The final outdoor sowing should be made at least eight weeks before the first frost is due.

Late-summer sowings of headed types can be made for cloching in autumn or for transplanting the modules under cover. These may not heart-up but will be tender enough for salad leaves and they will re-sprout. If the plants survive until spring, the flower shoots are edible, being tender and tasty. For these late sowings, use the most cold-tolerant cylindrical varieties, although barrel types will yield reasonable results. In a good season, leaves can be harvested from October to May. They should be spaced as for the outdoor headed crop or at 120–150mm (5–6in) for a cut-and-come-again crop.

Loose-headed types can be sown in early spring until early autumn, although very hot weather should be avoided. The last sowing in late summer or early autumn may need covering with cloches or fleece, if it is not transplanted undercover. The undercover crop should be spaced at 200mm (8in) for a semi-mature crop or 300mm (12in) for full maturity. With temperatures between 4–6°C the first cuttings can be made within 50 days of planting, with maturity at 80–100 days.

Soil Fertility and Rotation: treat them as other brassicas (*see* Chapter 3 for overview).

Irrigation: Chinese cabbage is generally very intolerant of drought (as they are shallow rooted) and are therefore a very thirsty crop.

Weed Control: dependent on the situation in which they are grown and the crop that is being cut. *See also* weeding in protected cropping and lettuce.

Pests and Diseases: Chinese cabbages are more susceptible to the common brassica pests and diseases than most orientals. Fluffy types are more resistant to diseases, although show greater susceptibility to flea beetle, slug and caterpillar attack. Where there is a high risk of flea beetle damage, particularly for summer sowings outside, fleece or mesh covers should be used.

Harvest and Storage: on average, the crop is ready for harvest two to two and a half months after sowing, although actual times vary from 55–100 days. Mid-July sowings mature fastest. Heads can be harvested when they feel solid and do not cave in when pressed firmly on the sides. It is best to store Chinese cabbage at 1°C with about 85 per cent humidity, when mature heads will keep for two to three months. However, they can keep for almost as long if pulled whole and stacked on their sides, five to ten plants high, in heaps with roots pointing outwards, then covered with sacking, either in an open shed or in the field. Loose-headed types can be harvested at any stage, including the flowering shoots.

Oriental Mustards (Br, *Brassica juncea* (various varieties))

This group of crops is gaining popularity as a salad leaf, particularly during winter months, as they are often very hardy. They have medium to large, sometimes red leaves and a distinctive, hot flavour that may intensify with maturity.

Oriental mustards

Types and Varieties: there is a vast range of oriental mustard varieties, which can loosely be divided into leafy types, curly types and swollen-stem types. Some of the more popular are detailed below.

Large-leaved group (Br, B. juncea *var.* rugosa*)*. These have large, sometimes red or purple-veined and edged leaves and a very hot flavour. They are very hardy. The stems are also edible and succulent.

Green-In-The-Snow group (Br, B. juncea *var.* multiceps*)*: these have medium-sized serrated leaves and a hot flavour. They are very frost hardy, as their name implies, and also show high resistance to pests and diseases.

Common group (Br, B. juncea *var.* foliosa*)*: these have coarse, dark-green, serrated leaves that are sometimes feathery, sometimes rounded. They are quite hardy but bolt rapidly in spring. The flavour is often milder than other mustards, although they can become hot if growth is checked and the flavour, therefore, concentrated.

Sowing, Planting and Growing: the seed of all these groups is small, so must be sown shallowly. They are best raised from transplants with two to three seeds in each module or block. Early crops of this cold-weather variety tend to bolt, so delay the first sowing until midsummer for mature plants that will stand or as cut-and-come-again leaves until the following spring, although re-growth from cut-and-come-again crops may be slow. For a baby-leaf crop the sowing can be later. This crop does very well undercover and only stops growing if the weather becomes very cold, although it will soon start growing again in late winter. Protected crops are sown in late summer or early autumn and can also be sown later for a baby-leaf crop. If seed is direct-drilled outdoors as soon as soil is workable in spring, it is possible to harvest baby-leaf crops during summer months, although these plants will bolt quickly if left to mature.

Spacing varies from 100–150mm (4–6in) for a baby-leaf crop to up to 450mm (18in) for mature plants of spreading varieties, with 300mm (12in) the average.

Soil Fertility and Rotation: as for Chinese cabbage.

Irrigation: the crop must not be allowed to dry out although, as the plants are very deep rooting, they do not require much irrigation.

Weeds: drilled crops will need a clean seedbed to establish well, otherwise weed management will be similar to other brassicas. An established and leafy crop will be able to outcompete other weeds.

Pests: flea beetles (*Phyllotreta* spp.) tend to be the most troublesome pests of these brassicas, although diamondback moths (*Plutella xylostella*) may also cause damage.

Diseases: a range of brassica diseases can potentially infect oriental mustards. They are probably best managed by following good cultural practices, as recommended for brassicas.

Harvest and Storage: as with most leafy salads it will wilt quickly once harvested and should not be cut when conditions are hot and dry. It should be stored in perforated bags or containers with air holes, at 3°C with high humidity. However, it should be used as quickly as possible after cutting and will not generally be saleable more than about five days after cutting.

Pak Choi (Br, *Brassica rapa* var. *chinensis*)

Pak Choi generally has thick, juicy, white stems (although these can be green) and large, rounded, dark-green leaves that can have a slightly 'savoyed' appearance. The flavour is mild. All parts of the plant can be eaten and are very succulent, including the flowering stems, although they may become stringy in very hot or dry conditions. Pak choi can be cooked but is also good raw in salads. They are not cold-tolerant plants but do well undercover over winter, which is when they are most useful for salad packs.

Types and Varieties: the common types are more or less sturdy with an upright growth habit and form a solid bulb-like base when mature. There are loose-headed varieties available also, with spoon-shaped leaves that form a prostrate rosette. The size of the plants varies from 100mm to about 450mm (4–18in) high, although they can reach 600mm (24in). The green-stemmed varieties of upright Pak Choi are more robust than the white-stemmed varieties, so perform better over the summer. The squat 'Canton' types are also more suited to warm weather, although they bolt more quickly when it gets cold. Loose-headed varieties are much more frost-tolerant than the more common upright types and are also slower growing and less vigorous, and therefore less nutrient-hungry.

Sowing, Planting and Growing: for mature heads, seed is sown in modules and later transplanted. Spacing varies, depending on the size of the variety, from about 150mm (6in) for short varieties to about 200mm (8in) for medium-sized varieties and 300–450mm (12–18in) for the larger varieties. Loose-headed types are spaced at 150–300mm (6–12in), depending on the size of the plant required. Pak Choi can also be grown as a cut-and-come-again seedling crop, in which case it is best to direct-drill.

As Pak Choi is very closely related to Chinese cabbage it is cultivated in much the same way. The earliest sowings should be made with heat, as the seeds need a minimum temperature of 18°C to germinate. Bolt-resistant and, if possible, cold-tolerant varieties should be used. For midsummer sowings the best results come from squat varieties, preferably heat-tolerant ones, and they should be harvested as soon as possible. The loose-headed varieties require a slightly longer growing period, so are sown mid- to late-summer.

Pak Choi can be grown as an intercrop, for example among sweetcorn or bean plants, because they have a very short growing season (six to eight weeks) and in summer the Pak Choi benefits from some shade. Cut-and-come-again crops can be grown in strips between rows of larger plants, such as cauliflowers or calabrese.

Soil Fertility and Rotation: a rich well-drained soil in a cool, but not too shady, spot. It tolerates a range of pH but prefers a slightly alkaline soil (pH 6–7.5).

Irrigation: it grows best when regularly watered, as it is shallow rooted and intolerant of drought.

Weeds: as for other brassicas, the ground needs to be well cultivated to allow the crops to establish in weed-free conditions, especially with direct-drilled crops, after which they compete against weeds.

Pests: as with all oriental salads, flea beetles (*Phyllotreta* spp.) tend to be the most damaging pest because of the shot holes eaten in the leaves. Floating covers (fleece or mesh) can help to establish the crop and later help to keep the pests off the crop.

Diseases: are not generally troublesome.

Harvest and Storage: mature heads can be ready in six weeks for small varieties, although eight is more common. The stumps will re-sprout if the whole head is harvested and individual leaves can also be taken. Cut-and-come-again seedlings can be ready in three weeks from sowing and may give up to three cuts.

Other Leafy Salad Crops

Amaranth (also Vegetable Amaranth, Leaf Amaranth) (O, Amaranthus *spp.)*: ultimately these bushy plants are harvested for grain, but they can be used at any growth stage, including being cooked like spinach. They require a fairly warm climate (20–25°C), therefore drill or transplant into warm soil (ideally above 20°C) in late spring, after all danger of frost is passed. The average spacing is 100mm (4in) apart each way. Amaranths perform best in light, fertile, well-drained soil and will tolerate some drought and a fairly acid soil. The red- and green-leaved forms are the most prolific and attractive in salad packs, but the white-leaved varieties have a better flavour. It is harvested as cut-and-come-again leaves from the seedling stage, with young leaves and shoots ready a few weeks from sowing. Individual leaves from more mature plants are ready for harvesting about fifty days from sowing in summer. Pick regularly and remove any flowers that develop.

Basil (O, Ocimum minimum*)*: is a tender perennial herb grown as an annual crop that has huge sales-potential and can be a top earner for protected cropping situations. It is good for the restaurant trade and for direct marketing and is best grown in a greenhouse or polytunnel, as it thrives in warm conditions. It needs at least 13°C to germinate and should be raised in modules, for planting in May or June 150–250mm (6–10in) apart. Basil is highly suitable for the outer beds of a polytunnel, as it is relatively low-growing (300–450mm (12–18in)) or it makes a good companion plant to go underneath tomatoes. The leaves need to be picked regularly to encourage bushy growth, ideally picking back to a node. Flowers should also be removed. Pick in the cool part of the day, cold store and pack into small, plastic bags for freshness. It can succumb to fungal diseases, such as *Botrytis*, later in the autumn under cold conditions and will be cut down by the first frost.

Beets (Ch, Beta vulgaris *var.* cicla*)*: the beets include leaf beet or perpetual spinach and the various forms of Swiss chard. There are some particularly colourful chards in the market, which make a salad pack look exciting. However, these types should be used in moderation, due to their bitter, slightly tough leaves. For cultivation, pest and disease information *see* the section on beetroot below. For baby-leaf crops the spacing can be reduced to 20–30mm (0.75–1.25in). They are very prolific plants, re-growing over much of the year to provide many cuts. Perpetual spinach is useful, as annual spinach is prone to bolt, due to dry conditions (*see also* beetroot below).

Chervil (Um, Anthriscus cerefolium*)*: has a faintly aniseed flavour and attractive fern-like leaves. It grows to about 300mm (12in) high, although it can be cropped when much smaller. It is very hardy and may stay green the whole winter, making it very useful for salad packs in this season. Modules can be sown and transplanted or seed direct-drilled and thinned to about 100mm (4in). Sow in spring under light shade for summer harvesting or in late summer for cropping from autumn onwards. Winter protection will provide a higher quality crop. If plants are left to run to seed, it will self-seed, so a permanent bed could be established. Chervil tolerates a variety of soil types and is not usually troubled with pests or diseases.

Coriander (Um, Coriandrum sativum*)*: a popular annual crop that can be fitted into a vegetable system as a catch crop. A sunny location is ideal, with light but fertile soil. It is relatively hardy and can be grown successfully outside until late autumn. Choose varieties that are suited for leaf production and sow regularly for a succession, drilling every ten to fourteen days from early spring to September. Irrigation can be useful and coriander needs no check to growth, as it can run to seed very quickly, especially in hot weather. The smell of bunched coriander can be major attractant for customers on a market stall. It can be drilled between brassica and other vegetable crops and any unharvested coriander can be left to flower to attract hoverflies for aphid control. Used in moderation, this herb's pungent flavour blends well with a mixture of salad leaves.

Corn Salad or Lambs Lettuce (O, Valerianella locusta*)*: there are large-leaved varieties (English or Dutch) of corn salad, which are the more prolific, although the smaller-leaved (French) 'verte' types are hardier. The main interest of this crop is as a winter alternative to lettuce, as the flavour is mild and the leaves and plants are small and slow-growing but hardy and ready earlier in spring than most other salad leaves. Individual leaves can be harvested – or the whole plant cut when small – and corn salad will often self-seed, so it may be possible to establish a permanent bed. Seed can be multi-sown in modules or seed trays and transplanted, direct-drilled or broadcast from early spring to late summer with a midsummer sowing for the main winter crop. Space at or thin to about 100mm (4in) apart and keep the soil moist until the seed has germinated. Mature plants are ready about twelve weeks from sowing, although cut-and-come-again leaves can be taken once the plants are about 50mm (2in) high.

Dandelion (O, Taraxacum officinale*)*: cultivated strains of this plant have long been popular on the continent and are gaining ground here, including a red-ribbed variety, which is a welcome addition to salad packs. Cultivated strains are more prolific than wild varieties and have been selected for flavour, although older leaves still require blanching for about ten days to reduce bitterness. Dandelion should be sown in spring and early summer, either in modules, seed trays or direct. Final plant-spacing should be about 350mm (14in) apart. Dandelion is a hardy perennial and will, therefore, re-grow in spring; alternatively, in very cold areas, it can be lifted and forced, as for chicory. The only soil requirement is good drainage.

Edible Oil-Seed Rape (Br, Brassica rapa *var.* oleifera *and* utilis*)*: this is cultivated for its leaves and flowering shoots, although it will go on to produce edible oil. The only common variety on the market at the present is the purpose-bred Bouquet F1, with thick, succulent stems, coarse, crinkly leaves and the characteristic yellow flowers. Best harvested before the buds open, this crop tolerates light frost and does well undercover. Sow from early to late summer in modules or direct-drill with final spacings of 150–200mm (6–8in). The shoots can be cut from 100mm (4in) high, though tips are still tasty from much larger plants.

Florence Fennel (Um, Foeniculum vulgare *var.* dulce*)*: this type of fennel is usually grown for its bulbous aniseed-flavoured roots, but the fine, feathery leaves can also be included in mixed-salad packs. The crop is of limited demand, but is becoming increasingly popular with the foodies and will augment a market stall, box or farm shop by its appearance and smell. There may be local demand from the catering trade. It is not the easiest crop to grow and will bolt with abandon given half a chance. Early or late crops can be grown under protection. Sown from April to July in blocks or modules, the seed germinates between 10–30°C. Planting out is between May to July. Successional sowings every ten to fourteen days should provide continuity. Bolt-resistant varieties need to be chosen for the early production. It could be direct-drilled in June or July when the soil is warm. Light soils are preferred, with good moisture-holding capacity and reasonable fertility. The pH should be between 5.5 and 7.5. Fennel needs to be kept growing and irrigation needs to be available, otherwise bolting will occur. Weed control should be reasonably straightforward, using any opportunity for weed strikes prior to planting and inter-row weeding thereafter. Slugs can damage the bulbs, otherwise few pest problems arc likely to be encountered. Later plantings miss the main carrot fly generation and diseases are unlikely to cause problems. A short window of opportunity is available for harvesting before they start to bolt in summer, cutting when the size of a flattened tennis ball, though this is slightly longer in the autumn. They will not store.

Komatsuna or Mustard Spinach (Br, Brassica rapa *var.* komatsuna*)*: the komatsuna group of greens is the result of crossing various brassicas, which accounts for the variety of flavours, with more or less spinach, cabbage and mustard overtones, depending on cultivar. The plants are very vigorous and tend to have glossy, green leaves that can grow very large, although the smaller, young leaves are more tender and tasty raw. They are very productive undercover as an over-winter crop and are prolific as a cut-and-come-again crop. Generally more robust than other oriental brassicas and more tolerant of extremes of heat, cold and drought, they are also less prone to pest and disease and to bolting. Komatsuna can be eaten cooked as well as raw. It is cultivated in a similar manner to Chinese cabbage and should be sown monthly from early spring to late summer, with the first and last sowings under cover. It can be direct drilled at about 50mm (2in) for cut-and-come-again seedlings, or in modules and transplanted at about 300mm (12in), depending on the cultivar, for a mature crop. A sowing under cover in early autumn will do very well for cropping over winter and can be augmented very early in the following season by mid- to late-winter sowings, also under cover.

Landcress or American Landcress (Br, Barbarea verna*)*: this very hardy plant has a strong peppery flavour similar to that of watercress, for which it is often grown as a winter substitute. It forms a central rosette with long, lobed leaves. Landcress is good for planting in damp areas, as it has a very high water requirement, inadequate moisture causing bolting and very bitter and tough leaves. Sowings are usually made in summer for a winter crop, although spring sowings will provide leaves in summer. The minimum plant spacing is 150mm (6in). It will tolerate some shade if moisture is available and it bolts easily in hot weather. Plants protected over winter will be bigger and lusher. Flea beetle can be a problem during early growth stages, which is another reason for frequent irrigation. The first leaves are ready for harvest about two months after sowing and can be picked as individual leaves or the whole plant harvested as a cut-and-come-again crop.

Leaf Celery (Um, Apium graveolens*)*: this is a smaller and hardier plant than stem celery, with fine leaves, and is especially useful in winter months when stem celery is not available. It usually stays green over winter. The flavour becomes stronger when the plants start to go to seed. The main sowing is made in spring or early summer in modules and transplanted to 125mm (5in) apart. In very cold areas, a mid- to late-summer sowing can be transplanted undercover. Cut-and-come-again harvesting can begin as soon as the plants reach a height of about 125mm (5in). For general cultivation and other information *see* the section on celery.

Mizuna and Mibuna (Br, Brassica rapa *var.* nipposinica*)*: strictly speaking, these two are part of the oriental mustards group, however, the mustard flavour is so undetectable and they are so much better known than the others that they deserve individual mention. Mizuna forms clumps of fern-like leaves up to 300mm (12in) or more and has a mild lettuce-like flavour with a hint of mustard. Modern varieties have broader, less serrated leaves. Sow two to three seeds in each module at monthly intervals from early spring and plant at up to 250–300mm (10–12in) apart. A final sowing in late summer transplanted undercover does very well as a winter salad. Flea beetle can be a problem. Cut individual leaves from about 100mm (4in) or the whole plant as a very productive cut-and-come-again crop. Mibuna is very closely related to mizuna but has long, strap-like leaves without serration and a slightly stronger flavour. It is slightly less hardy.

Orache or Mountain Spinach (Ch, Atriplex hortensis*)*: orache is sometimes available as mixtures of seed, as the leaf colour includes a variety of reds, ranging from darkish brown to scarlet, as well as green. It has a slightly spinachy flavour and faintly hairy leaves, so only use young and tender leaves, either individually or as a cut-and-come-again crop. The plants like a fertile soil where they can reach up to 2m (6.5ft) in height. Generally, growing conditions are the same as for spinach. The seed is best direct-drilled. For mature plants, start outdoors in March and thin to 200mm (8in). The plants are very vigorous, therefore, the leaves need regular picking. Orache will self-seed for a permanent bed. For cut-and-come-again seedlings, sow successionally from late spring to late summer. An undercover sowing in late autumn may stand through the winter. Harvesting begins within six weeks.

Parsley (Um, Petroselinum crispum*)*: this herb is particularly useful towards the extreme ends of the season when other salad-pack material is becoming scarce. It is very hardy, lasting into early winter and re-growing again very early in spring. The curled varieties are more hardy and robust, but the flat leaf (French) varieties have a sweeter flavour and give a more attractive display.

Rocket, also Rucola, Salad Rocket, Roquette (Br, Eruca sativa *spp.* sativa*)*: rocket has a small, green, serrated leaf with a hot spicy flavour. It is best direct-drilled, though not too densely or the seedlings will bolt. Cool temperatures are ideal for growth and mature plants may over-winter, producing a flush of leaf late in the season. Even small plants are fairly hardy and will stand a light frost. The seed can be sown from early spring to midsummer for an outdoor crop or early autumn for an undercover crop, where it does well. Avoid midsummer sowings in hot areas or sow in light shade. It is very easy to save seed from the mature seedpods and seed remains viable for several years. Good irrigation or moisture-retentive soil is best. Flea beetle can be a problem in early growth stages, though early and late sowings can be protected, to some extent, with horticultural fleece or mesh crop covers. The most time- and cost-effective method of harvest is to cut whole plants from about 150mm (6in). These will re-grow and allow two or more further cuts. Alternatively, space plants at 150mm (6in) and allow them to mature, picking individual leaves as required.

Salad Rape (Br, Brassica napus*)*: mild flavoured and fast growing, this plant is often used in mustard and cress as the 'mustard' component. It is very hardy and responds well to cut-and-come-again treatment, giving three or sometimes four cuts, although if left it can grow to 600mm (2ft) tall and be cooked whole as greens. The flavour remains mild despite maturity. Salad rape bolts quickly in hot weather, but re-grows fast in spring from a late-autumn sowing, so is useful early in the season. It germinates at lower temperatures than many salad plants and can survive quite severe frosts. Direct-drill or broadcast in spring or late summer to autumn, with the earliest and latest sowings undercover. The first harvest can be made as soon as ten days after sowing. It is very susceptible to slugs.

Salsola (Ch, Salsola soda/Liscari sativa*)*: salsola is an easy-to-grow Asian salt-marsh plant, related to our marsh samphire. Several species are cultivated in Asia, although they all have the characteristic green, matchstick leaves, which can be up to 60mm (2.5in) long. It has a crunchy texture, salty flavour and can be eaten raw or cooked. The seed loses its viability quickly and is best stored before sowing at 5°C in winter and 5–15°C in summer. The first sowings can be made in spring and continue till early summer. It can be direct-drilled and then thinned to 60mm (2.5in). Salsola tolerates fairly poor soil and prefers alkaline conditions. Cutting can begin when the plants are only a few centimetres tall.

Shungiku, also Chrysanthemum Greens, Chop Suey Greens, Garland Chrysanthemum (O, Xanthophthalmum coronarium *or* Chrysanthemum coronarium*)*: as a mature plant, shungiku is usually cooked but it is good raw as a cut-and-come-again crop when young. The leaves can have the typical, feathery, chrysanthemum shape and a strong and distinctive taste. The flower petals are also edible, but not the bitter centre and, indeed, the leaves also become bitter if allowed to flower. The broad-leaved varieties are the most prolific and need to be picked regularly to keep them productive. Nip-out growing points if the plants become too leggy or woody. The plant is slightly hardy and will tolerate a few degrees of frost, as well as light shade and a wide range of soil types, although it prefers an acid soil. Sow in modules and transplant or sow thinly *in situ* thinning further to 125mm (5in) per plant. Shungiku runs to seed rapidly in hot weather but can be sown from mid-spring to early summer with early spring and late summer sowings made undercover. It can also be propagated from cuttings of side shoots before the plant runs to seed. Cropping starts when the seedlings are about 50–100mm (2–4in) high, about six weeks after sowing, and should give a few cuts before running to seed.

Spinach (Ch, Spinacea oleracea*)*: baby spinach leaves are often included in salad packs for their tender leaves and flavour. The thicker, more-puckered leaved varieties are used, as they are more robust and less inclined to bruise. They are usually direct-drilled and can be cut from about thirty to forty days after sowing, with as many as four cuts possible from each batch. They should be sown successionally, every three to four weeks, but bear in mind the plants may bolt quickly with very early sowing in hot weather.

Texsel (Texel) Greens (Br, Brassica carinata*)*: have nutritious, glossy leaves with a spinach-like flavour that can be eaten raw as a cut-and-come-again crop or can be cooked. The seed can be direct-drilled in rows 150mm (6in) apart from early spring undercover, with outdoor drillings at three-week intervals from mid-spring to late summer. Late autumn drillings should be made undercover, as the plant is only moderately hardy. The plants mature very quickly, so can be sown into clubroot-infested ground and harvested before infection takes hold.

Winter Purslane, also Claytonia, Miner's Lettuce (O, Montia perfoliata*)*: winter purslane has slightly succulent, heart-shaped leaves, a very mild flavour and high vitamin C content. The stems and flowers are also edible and it is fairly hardy. It is best grown undercover over winter from a late-summer sowing. It likes a well-drained soil, but otherwise tolerates a wide range of conditions and, although an annual plant, self-seeds freely once established. Seed is very small, so direct-drilling is easiest, although it can be multi-sown in modules and planted at 150mm (6in) in the row, where it provides a higher yield. Winter purslane can be grown almost all year and is sown in spring for a summer crop, however, it performs best as a late autumn and early spring crop sown in summer. It will stop growing in midwinter but start again in early spring. Undercover slug predation on drilled seedlings can wipe out an entire batch, although no other pest or disease problems are common. The first cut can be made as soon as eight weeks after sowing and a cut-and-come-again crop should provide at least two cuts, although single leaves can be harvested from mature plants or the whole head taken.

LEAF AND FLOWER VEGETABLES

Leaf and flower vegetables are mainly represented by the brassicas in field vegetable production. Of all the vegetable crop groups, the brassicas are the most diverse and possibly the most widely cultivated in the UK and Europe. They originate from the Atlantic and Mediterranean sea-coasts of Europe and so they are more or less well adapted for growing in the UK climate. Many can be produced all year round (depending on type and variety). Leaf and flower brassica types include the most commonly cooked vegetables, like Brussels sprouts, cabbage, calabrese and cauliflower. Surprisingly, all these crops have been developed from one species: *Brassica oleracea.* Brassica crops share many characteristics and many of these have been referred to in Chapter 3, which should be consulted for more specific details on topics such as soil fertility, irrigation, pests, diseases and weeds. Such topics are only listed within the individual-vegetable headings below where the information is specific to that vegetable type. Leaf beets and spinach are also popular leaf crops and these are also discussed.

Brussels Sprouts (Br, *Brassica oleracea* var. *gemmifera*)

With the advent of sweeter-tasting varieties, Brussels sprouts have gained in popularity over the last few years, although the Christmas period is still the most important marketing period for this vegetable. They remain in the ground for an extended period and are an over-wintering crop that can be harvested from October until February, depending on variety. Popular earlier varieties include Diablo and Maximus, which are harvested from October to January, and later varieties Doric and Revenge, harvested between December and February.

Sowing, Planting and Growing: most crops are sown under cover in March, then planted out in April or May at a spacing of 600mm (24in) between plants and 600mm (24in) between rows. Many commercial growers also top the plants, to encourage button development.

Soil Fertility and Rotation: Brussels sprouts require firmer soil than most brassicas, as plants can fall over in wet and windy conditions, spoiling the button quality. They have a long period of growth and a long main root, so need a deeply worked soil and are not at all tolerant of poor soil structure. Sub-soiling may be necessary. They prefer heavier loams and need a good supply of nutrients. Too much nitrogen can cause 'blown sprouts', however, as can a loose and puffy soil.

Irrigation: similar to other brassicas (*see* Chapter 3 for an overview).

Weeds: similar to other brassicas.

Pests: the long period between planting and harvesting makes this crop vulnerable to attack by an array of pests and diseases, making it a difficult crop to grow organically. Infestation by cabbage aphid (*Brevicoryne brassicae*) from September onwards can seriously spoil the appearance and can build up to epidemic levels under mild conditions.

Diseases: dark leaf-spot and ring spot can also spoil the appearance of the sprouts and tend to be more serious in this crop than other brassica types. In variety testing programmes some varieties have been selected as being more resistant to these diseases. The variety Revenge, for example, has good resistance to these diseases, according to NIAB ratings, but these are liable to change and should be consulted on a regular basis.

Harvest and Storage: a yellowing of the lower leaves indicates the first sprouts are ready. Harvest can start with early varieties in early September, with maincrop from October, with the flavour usually considered better after the first frosts. Early season sprouts can be sold on the stalk if the appearance is consistently good, saving considerable labour costs in cutting. These can be popular in farm shops or on market stalls but are less suitable for boxes, due to the bulk. Later sprouts will generally need to be picked by hand, with a quick downward-pressure of the thumb, removing lower leaves as you go. This is not a popular job on freezing December mornings and washing-up gloves are advised! Minimum size should be 15mm (0.6in) for untrimmed sprouts or 10mm (0.4in) for trimmed. Trimmed sprouts need to be a good colour and relatively free from blemishes. Some degree of discoloration and harvesting damage is tolerated on the outer leaves of untrimmed sprouts.

Cabbage (Br, *Brassica oleracea*)

There is a vast array of cabbage types and varieties with a range of times of maturity, so that the market can be supplied all the year round. Summer types can be sown from February to March and harvested from June to November. They are not frost hardy and do not stand well, so must be sold soon after maturing. It is important when planning a succession of varieties to ensure that the market will not be oversupplied, as there is often a drop in the demand for vegetables such as cabbage during the summer months. Pointed cabbage types can also be harvested earlier, as spring greens before they have produced a heart, if leaves do not suffer

Loose Brussels sprouts after picking

from too much pigeon damage. This is often a time of year when there are few green vegetables available in the UK market, so they can sell well. Red or white cabbage is harvested from July to October. These are also sensitive to frost damage, so must be harvested before the winter, but they have good storage potential for providing supplies over the winter. Winter cabbages are available as January King types, Savoy or hybrid Tundra types. All can remain in the field over the winter and can be harvested from October until March.

A range of cultivars of each of these cabbage types is available from seed catalogues and many have been evaluated under the NIAB organic variety-testing programme, so descriptions are available as to their performance.

Sowing, Planting and Growing: spacings of cabbage can range from 400mm (16in) each way for fast-growing summer types, with the spacing widened up to 700mm (28in) in the row for slower-growing winter types (*see* Table 12.1). Spacing should aim to satisfy the target size for the market. Generally, consumers favour smaller-sized cabbages, especially those in box schemes, so closer spacings now tend to be more popular. Pointed cabbage can be planted at very close spacings – 200mm (8in) – and half the crop removed early for spring greens, whilst letting the remainder of the crop produce hearts.

Table 12.1 Suggested sowing, drilling and transplanting dates for winter cabbage

Crop	Frames or cold glass	Outdoor seedbed	Transplanting completed by
Winter white for storage	3rd week April	early to mid-April	3rd week June
Savoy, January King	early to mid-June	mid- to late-May	3rd week of July
Winter hybrid	mid-June	late-May to mid-June	end of July

Soil Fertility and Rotation: similar to other brassica types (*see* Chapter 3 for an overview).

Irrigation: similar requirements to other brassica types.

Weeds: similar field management to other brassica types.

Pests: cabbages are troubled by the same pests as other brassicas. Cabbage aphids (*Brevicoryne brassicae*) are more likely to cause problems in crops harvested after September, when numbers peak. They can be a

Pointed-cabbage types can be harvested earlier in the season

particular problem in Savoys, where the debris becomes trapped in the crenulations of the leaves. Specific control-measures have been discussed in Chapter 3 but novel methods, such as intercropping, might also be beneficial in reducing aphid infestation in some cases.

Diseases: cabbages generally suffer the same diseases as other brassicas. Dark leaf-spot (*Alternaria brassicae*), ring spot (*Mycosphaerella brassicicola*) and black rot (*Xanthomonas campestris*) tend to be more of a problem in over-wintered cabbage that remains in cold, damp conditions.

Harvest and Storage: for cabbages harvested in the field for immediate sales, outer leaves can be removed, but if too many leaves have to be stripped back, not only is there a reduction in yield, but the head can be rejected for being too pale in colour. It is preferable to leave one layer of outer leaves for protection. Winter white and red cabbages are not able to withstand winter frosts and should be harvested and in store by mid-November. Handle the cabbages carefully into clean boxes, avoiding bruising. They can be stacked in boxes or crates in a barn and should store until the end of March, with the heads trimmed down as they come out of storage. For storage until April and through to July, cold storage is required.

Calabrese/Broccoli (Br, *Brassica oleracea* var. *italica*)

Calabrese is a popular brassica. A range of cultivars has been evaluated under organic conditions and is available as organic seed. Currently, Marathon is by far the most widely commercially grown variety, producing a large, tightly beaded head of attractive appearance. For wholesale or packer outlets, calabrese is normally grown just for the large, central head but, for direct marketing, the side shoots, which develop later, can be harvested over a longer period. New varieties are always being produced and growers should keep an eye on the results of NIAB and seed company testing programmes, whose results are frequently reported in the growing press.

Sowing, Planting and Growing: planting should be timed so that there is little risk of exposing heads to frost. Earlier crops may be grown under fleece, to take advantage of the early market. Modules can be sown from March onwards under cover and will be ready to plant out after four to six weeks. Crops should be ready to harvest sixty to eighty days after planting. Successional sowings can produce a crop from June into November, depending on the risk of frost in the area. Spacing can range from 300–600mm (12–24in) between rows with 300–450mm (12–18in) between plants within the row. Crops aiming for an early harvest of a single-head per plant should be planted at a closer spacing, whereas, if side shoots are to be cut, a wider spacing is needed.

Soil Fertility and Rotation: calabrese is a particularly nutrient-hungry crop and requires higher fertility levels than other brassicas, otherwise heads tend to be small and of poor quality. The crop should be carefully sited in the rotation, either after a fertility-building period or manure application.

Irrigation: adequate moisture levels are particularly important during head formation, to ensure good yield and quality.

Weeds: like other brassicas, early weed management is necessary until the crop grows.

Pests: caterpillars and their droppings can be a particular problem close to the point of harvest, especially when the crop is destined for supermarkets, which have very low tolerance for pest levels in the finished product. Sometimes a Bt (*Bacillus thurengiensis*) spray when caterpillars are small may be necessary to achieve the low levels required. Similarly, flea and pollen beetles have occasionally been a problem in this crop, as they are very difficult to remove from the heads.

Diseases: are not usually a problem with this brassica as, although leaves often display a range of disease symptoms, they are not harvested and sold.

Harvest and Storage: where the crop is sold as individual heads, they need to be cut whilst the beads still have a tight, compact appearance. Crops sold as florets can be harvested when the heads are slightly looser, to achieve extra yield. After the main head has been harvested, plants will continue to produce side shoots, which can be taken for florets.

Cauliflower (Br, *Brassica oleracea* var. *Botrytis*)

The edible portion of the cauliflower is the curd, which is an arrested stage of flower development. Cauliflowers can be divided broadly into three maturity groups: summer, autumn and winter. There is a wide range of varieties from which to choose, many available as organic seed. Some variety trialling work has been done by NIAB and seed companies hold regular open days where varietal performance is discussed.

Types and Varieties: summer varieties, including Cassius, Fargo and Fremont, are sown in February to May, planted out in April to July, then harvested from July to October. Autumn varieties, such as Belot, Pierot and Maginot, are planted out after July and harvested in November. Winter varieties are also planted out after July and grow slowly during the winter for cutting from December until March. A range of varieties that mature at different times can be planted, to give a successive harvest over the winter. There is also an increasing demand for novelty types, such as Purple Cape (which heads in February and March when other brassicas are in short supply) and green or Romanesco types, sold by one Lincolnshire grower as

Cathedral Cauliflowers! These can add colour and interest to a box or market stall.

Sowing, Planting and Growing: plant spacing ranges from 500mm (20in) each way for the faster-growing summer types to 700mm (28in) each way for the slower-growing winter types.

Soil Fertility and Rotation: as with calabrese, cauliflower requires high levels of fertility to ensure good quality curd formation and should be sited at the appropriate point in the rotation.

Irrigation: adequate moisture levels are essential for good quality curd-formation. However, overhead irrigation when the curd is opened should be avoided where possible, as this promotes the proliferation of fungi such as *Alternaria* on the curd.

Weeds: similar to other brassicas.

Pests: cauliflowers suffer from the same range of pests as other brassicas. As with calabrese, caterpillars and their droppings can spoil the appearance of the head.

Diseases: cauliflowers suffer from the same range of diseases as other brassicas. Over-winter cauliflowers can be more susceptible to ring spot, especially if conditions are wet. In serious cases, this can defoliate the crop. *Alternaria* or dark leaf-spot can make the crop unsellable if it gets onto the curd causing black blemishes; in such cases it may be necessary to harvest before the curd reaches full size. Hollowing and internal browning where the stem joins the curd can be a result of irregular watering or boron deficiency.

Harvest and Storage: as with calabrese, the curds should be harvested when they are still tight and compact, if they are to be sold as whole heads. A few outer leaves should be left on to protect it from damage, though these are usually trimmed down to a few centimetres above the curd.

Other Leaf and Flower Crops

Chard or Swiss chard (Ch, Beta vulgaris *var.* cicla*)*: not quite as fast growing as leaf spinach, chard makes a very good cut-and-come-again crop and will stand through the winter to re-grow in spring. Some chard varieties have extremely brightly coloured stems, although the commonest Swiss chard has white leaf-ribs and stalks. On other varieties, the green leaf is contrasted with all shades of red, pink, yellow and orange stems and ribs. It is possible to buy a colour mix, usually called 'Bright Lights' or 'Rainbow'. The leaves of chard are tougher and coarser than leaf beet or spinach, although they have a higher iron-content. The stem is often very broad at the base, so that it is common practice to cook stems and leaves separately. Growing requirements are very similar to beetroot (*see* below), to which these leaf crops are closely related.

Harvested and trimmed cauliflower-heads

Kale (Br, Brassica oleracea *var.* acephala*)*: kale or borecole is not a very popular crop, but is a good source of winter greens. A range of leaf types exist, including curly leaved, black kale (palm cabbage) and red leaved, that add variety and colour to box schemes or market stalls during the winter months. More unusual varieties, such as Nero di Toscana and Red Russian, may appeal to this market. It is usually sown in modules from May to July and transplanted from July onwards at a spacing of 600mm (24in) each way. Kale is a frost-hardy crop that can be harvested from autumn and into winter, depending on the variety. Individual leaves can be pulled off, leaving the hearts to grow on for a second cut.

Leaf beet (Ch, Beta vulgaris *var.* cicla*)*: is very similar to the chards. It is often sold as 'spinach', especially at farmers' markets or through box schemes, as it has a similar – though less delicate and distinct – flavour and it can be used in exactly the same way; it is much easier to grow. Many people are unaware of the distinction between the two plants and the names 'leaf beet' or even 'perpetual spinach' are not commonly known. The leaves are coarser than true spinach and more robust, which makes harvest and storage easier, and the plant is more tolerant of heat, cold and drought, as well as more disease resistant. As for chard, leaf beet is a good alternative source of winter greens to the brassica family, when standards preclude the planting of that family. The growing requirements for this crop are very similar to beetroot, to which it is closely related (*see* below), although the leafy members of the beet family perform better when sown under cover in early autumn and early spring for winter and spring crops. The crops are not generally damaged to any great extent by pests, although slugs and birds may be a local problem. Foliar diseases can be more troublesome, as the leaves are sold. Common diseases are downy mildew (*Peronospora farinose*) and leaf spot (*Ramularia beticola*). They are closely related diseases that persist on plant debris, which is one good reason for maintaining a long rotation, and pass from crop to crop in moist conditions. There is also a risk of manganese deficiency, especially on poorly drained, highly organic soils and where pH is high, so care must be taken not to over-lime.

Purple-sprouting broccoli (Br, Brassica oleracea *var.* italica*)*: this traditional vegetable has experienced a resurgence in popularity in recent years. It has the advantage of being cold tolerant, so can provide broccoli spears during the winter months as an alternative to importing calabrese. Standard varieties are sown from April to June, transplanted in July and harvested between December and March (depending on variety). Plants are spaced at 600mm (24in) each way to allow side-shoot formation. After cutting the main head, the plants will provide a continuous supply of spears from side shoots. There are also varieties, such as Bordeaux, that require no cold period and will produce sprouting broccoli from June to November, although this will be competing against the main calabrese-market. They are not usually troubled with too many pest and disease problems, although pigeons can be a problem during crop establishment and in winter and early spring, when they will also attack the tops, as there is little alternative food available to them. Shoots can be cut or snapped off and will cold-store for up to a week in perforated bags or covered crates, although the leafy stalks will wilt if mishandled.

Spinach (Ch, Spinacea oleracea*)*: there are three main types of spinach: savoy, semi-savoy and flat-leaf. Baby-leaf spinach is regarded as a salad-leaf type (*see* above). Savoy spinach has crinkly, dark-green curly leaves. Flat-leaf or smooth-leaf spinach is unwrinkled and has spade-shaped leaves that are easier to clean than the curly types. The stalks are usually very narrow and tasty. Semi-savoy is a mix of the savoy and flat-leaf. Spinach is not usually classified by variety, but according to sowing time (spring, summer, and winter spinach) and harvesting method. Leaf is harvested by hand and root spinach is done by machine. The boundary between summer spinach and the autumn types is blurred, but both types can be eaten raw, although autumn spinach is usually tougher. Winter spinach has the most robust and strong-tasting leaves, which are coarser and often curly. Some foliar diseases may appear but are not usually serious. Spinach leaf-spot (*Cladosporium variabile*) is common and widespread in spinach crops, but is not usually a serious problem. The other most common disease of spinach is downy mildew (*Peronospora farinose*).

FRUIT VEGETABLES

Fruit vegetables are delicate and, therefore, easily damaged and difficult to handle. They cover a range of crop types and plant families and many of them are best produced as protected crops. However, they often provide a welcome splash of colour to a vegetable selection and are not difficult to grow when provided with the right growing conditions.

Aubergines (Sol, *Solanum melongena*)

Aubergines are a good crop for greenhouse and tunnel production and by no means as difficult to grow as some books would suggest. They can be treated very similarly to tomatoes and come in a wide range of shapes and colours, with the deep-purple oval types being most popular. Varieties can vary a lot, but F1 hybrids tend to grow more uniformly and produce higher marketable yields. The market tends to prefer shiny-purple/black fruit with green calyxes, which contrast well with the fruit. Other types include white egg-shaped fruits (hence the name Egg Plant in the USA), long, slender, finger-shaped fruit and the small, round, green or mottled types popular in Asian cookery. As Thai cooking is growing in popularity, there may be potential for these more unusual versions in the future, both for domestic and restaurant use.

Sowing, Planting and Growing: aubergines are sown in seed trays or modules between January and March when temperatures are at least 18°C, eight to ten weeks before they are to be planted out. Sometimes it is necessary to transfer the small plants to larger pots. The crop should be planted-up from April onwards, when the danger of frost is past, using a double row or bed with 450–550mm (18–20in) between plants. The plants can be supported with a 1–1.2m (3.25–4ft) stake or with hanging strings. Pruning is recommended for high-quality fruit – selecting two or three stems and pinching out the side shoots (as in tomatoes). The older, lower leaves should also be removed, to allow for more air circulation and light penetration. For polytunnel or cold-glass production, allow six to nine fruits to develop per plant. For long-season heated-glass production, a single stem can be trained, as with tomatoes, and more fruit can be set. Aubergines thrive on higher-day and lower-night temperatures and dislike high humidity, so the ventilation should be manipulated accordingly.

Soil Fertility and Rotation: aubergines have similar nutritional requirements to tomatoes and thrive in deep, well-structured loamy soils. The ideal pH is 6.5.

Irrigation: drip or trickle irrigation is best, as poor fruit set can be associated with high humidity.

Weed Control: control weeds as for other tunnel crops, through mulching, under-sowing or hand weeding.

Pests: are similar to tomatoes, with red spider mite and aphids likely to be the main concern; regular monitoring is essential and biological controls can be introduced as necessary.

Diseases: are similar to tomatoes but not normally a problem on this crop.

Harvesting and Storage: the fruit should be harvested while the skin is shiny, using a sharp knife, leaving the calyx attached to the fruit. Plants can be picked-over twice weekly. They should be handled carefully, as the

Harvesting aubergines grown in a polytunnel

skin can be easily damaged at this stage. They can also be subject to chilling injury when stored at temperatures below 7°C.

Peppers (Sol, *Capsicum annuum*)

There are a large number of shapes, colours, types and varieties of peppers. They are a tropical plant that needs to grow under protection, requiring warmer conditions than tomatoes. The two main types are chilli peppers and sweet peppers. Chilli peppers are probably best considered a specialist market, perhaps for direct sales. The plants don't tend to be as vigorous or as high yielding as the sweet peppers. In contrast, sweet peppers have become almost commonplace and an increasing variety of colours is now available. Although all types are green in the immature stage, different types will ripen to orange, yellow, purple, red and shades in between. Green peppers are the easiest to grow and the highest yielding, as production is slowed when fruits are allowed to ripen, which seems to take an age. The details given below apply mainly to sweet peppers. Details on chilli peppers are best found in more specialist texts.

Sowing, Planting and Growing: seed can be sown into trays at high temperatures (24–28°C) and pricked out into pots or modules. Germination can be slow and erratic for the open-pollinated varieties. For a heated-glasshouse crop, sowing can start from November to March, planting out from December to April. For a cold-glass or tunnel crop, sow in March/April for May planting. High and constant temperatures of 22–24°C should be maintained if possible, though they will grow slowly at lower temperatures. The crop should be planted out at 400–500mm (16–20in) apart in beds (although chilli peppers can be planted closer together). Stem density is more important than plant density and 6–8 stems/sq m is normally satisfactory. Two stems are normally trained per plant and any lateral stems are stopped at the one- to three-leaf stage. Strings attached to a top wire or similar supports are often used to hold the stems and pruning is necessary in order to prevent odd-shaped fruits developing. The plants are sensitive to cold weather and growth is inhibited below 10°C, whilst pollination can be poor below 15°C and above 32°C. Peppers like high humidity and can be grown with cucumbers, which have similar requirements in a mixed tunnel.

Soil Fertility and Rotation: as for tomatoes.

Irrigation: overhead irrigation should be avoided, as it can encourage grey mould (*Botrytis* rot). It is also best to avoid watering at night, which can encourage splitting of fruit. Even-application of watering will reduce the risk of blossom-end rot, which is mainly a physiological disease.

Weed control: as for other protected crops.

Pests: mice and slugs are a real problem for the ripening fruit (of sweet peppers), as they start to feast on them as soon as the peppers start to colour-up and become sweet. It might be necessary to harvest early and continue to ripen in store, if they are a serious problem and the services of a cat are not available. Aphids can be a problem, especially on the young leaves. Biological controls can be introduced and natural predators should be encouraged.

Diseases: resistant varieties are available against tomato mosaic virus. Damping-off problems caused by fungal diseases can occur where drainage is poor. *Pythium* can be a problem at low temperatures, *Rhizoctonia* at higher temperatures. Grey mould (*Botrytis cinerea*) can thrive under very humid conditions on leaves, fruit, flowers and stems (especially in wounds caused by management operations). Keep plants dry, particularly at night and avoid overhead irrigation.

Harvesting: the green-fruit stage is achieved five to nine weeks after flowering and fruits normally take another two to four weeks to ripen to colour. Harvest with a small, sharp knife, being careful not to damage the fruit. A rough wound on the calyx can cause *Botrytis* grey mould to set in. It is best to harvest the fruit in the morning before it has warmed up, and plants need to be picked over weekly. At the end of the season, if frost threatens, chilli-pepper plants can be pulled and hung upside down in a frost-free shed, if space permits.

Winter and Summer Squashes (Cuc, *Cucurbita pepo, C. maxima, C. moschata*)

Squashes derive from the cucurbita or Cucurbitaceae family of plants. They have their origins in South America and were amongst the first domesticated vegetables. Types include winter squashes (such as pumpkins and gourds) and summer squashes (including courgettes and marrows). Courgettes have become more popular since the 1960s, with changing tastes and more Mediterranean-style diets, and are now available all the year round as imports. Squashes, with their culinary versatility, are also increasing in popularity and are very useful for box schemes, as they can be stored easily and extend the range of choice and season. Recipe suggestions included in farm newsletters can help consumers discover these under-rated vegetables.

Types and varieties: the main types and varieties of squashes are briefly described below.

Courgettes and summer squashes: these are all members of the cucurbit species *Cucurbita pepo*. Summer squashes tend to be determinate or bush types and can be planted at 600mm (24in) between the plants on a bed, allowing wider rows for picking. They are used in the immature stage.

Courgettes can be prolific croppers. It is important to get them early in order to get the best prices and to get a foot in the market. Prices tend to drop when they glut in August, which coincides with when many customers are on holiday. Pigs can be useful at this time to dispose of the surplus crop. A range of organic seed is available in catalogues, but Ambassador F1, a standard commercial variety, is not easy to obtain as organic seed. There is a small market for courgettes with flowers, particularly for restaurants. Yellow courgette varieties, such as Soleil and Goldrush, are useful for this restaurant trade, on farmers' markets and to provide colour in boxes, but tend to be lower yielding. Round varieties in green and yellow are also available.

Apart from courgettes, other summer squashes include patty pans and crooknecks, which, like courgettes, need to be picked when young. Marrows are less popular and sales generally limited, although specific marrow varieties, such as Zebra Cross or Tiger Cross, are available and suitable for UK growing conditions. Overgrown courgettes should not be sold as marrows. Although summer squashes can be grown under protection in polytunnels for an early crop, they can be difficult to manage, often producing abundant top growth and little fruit, which is, in any case, vulnerable to rots. They are probably only suitable in larger tunnels with good ventilation and air movement.

Winter Squashes: the majority of winter squashes are indeterminate in habit and will trail vigorously around the holding. For this reason they need to be spaced at least 900mm (3ft) apart. Most winter squashes are either *Cucurbita moschata* or *maxima* species. Others are members of the *pepo* species (for example acorn squash and spaghetti squash) but are not long storers and should not be kept for more than two months. The *moschata* category includes butternuts. They need more warmth than some squashes and can struggle to ripen in a poor season. Squashes in the *maxima* category have short, corky stems and include Hubbards, Kabocha or Buttercup, and Turks Turban, as well as some large pumpkins. The best varieties for commercial use are the smaller – 1–2kg (2–4.5lb) – fruiting types. Uchiki Kuri or onion squash has red fruits with golden flesh averaging 1.5kg (3.25lb) in weight and is a good, reliable performer and early cropper. The kabocha or buttercup types are also of a similar weight, with chestnutty flesh and great eating qualities. Other good, smaller squashes are Harlequin and Sweet Dumpling, both gold- and yellow-striped squashes that are suitable for stuffing. Gem squashes produce an early crop of tennis-ball-size fruit that bake well. The Turks Turban is a striking novelty crop that is very ornamental, but not great for cooking. Some of the larger Hubbards, Crown Prince and Queensland Blue squashes have great storage and taste, but their size

Courgettes ready for marketing

can restrict sales to all but the most committed customers. Blue Ballet is a smaller version of Blue Hubbard, which produces intermediate-size fruit that is a good compromise.

Pumpkins: are now more associated with Halloween than with winter food supplies. Despite the fact that customers may not eat them, they can be a useful crop to bring new people in to farm shops or market stalls and are relatively easy to grow. Most pumpkins are of the *Cucurbita pepo* species and include the large jack-o-lantern and pie types. Some of the large types for showing are *Cucurbita maxima*. They are more vigorous than the winter squashes and need a wider spacing of 1,200mm (4ft) each way. They will not store much beyond Christmas and, in any case, you should aim to sell the majority by Halloween.

Seed and Raising Transplants: cucurbits can be either direct-drilled or transplanted from greenhouse-raised modules or blocks. Given the shortness of the UK season, winter squashes are better transplanted after the expected date of the last frost, usually towards the end of May. Courgettes and pumpkins can establish quickly from drilling if the soil temperatures are warm – and the soil needs to be at least 15°C. For field production of cucurbits, they should be sown under glass from the end of April, or one month before the expected planting date, keeping the trays out of the reach of mice, who love to eat the seeds. Take care to place the seeds on their sides in the cell or block, otherwise they can be vulnerable to rotting.

Cultivation and Growing: cucurbits can be planted through black-plastic mulch for weed control (biodegradable options are available), which will also advance earliness. Transplanting under fleece can also give an earlier crop. These mulches can be laid by machine, but planting will need to be done by hand through holes cut in the plastic. The paths between beds can be sown with a legume, such as clover or trefoil. If grown in the open field they can be machine planted. Another option is to plant or drill the crop into strips rotovated in an established clover-ley.

Soil Fertility and Rotation: squashes prefer a well-drained sandy loam with high organic matter and a pH of 6.0–6.5. Work the soil to a good tilth and incorporate manure or compost.

Irrigation: is usually necessary for establishment and to increase yield of summer squash. Drip or trickle irrigation can be employed under plastic mulches.

Weed control: mechanical cultivations and hand hoeing for weed control are possible until the trailing types start to run, so it is important to get the timing right. Cultivation should avoid injuring the shallow roots. Mulching, as described above, is widely practised.

Pests: losses from slugs can be high on heavier soils when planting through mulches. Mice and other rodents can be a problem, both for eating the seed and for the ripening crop. There are few other pests that bother cucurbits in the field in the UK.

Diseases: the biggest problem is powdery mildew (*Erysiphe cichoracearum*), with grey/white powdery patches colonizing leaves in the autumn, reducing photosynthetic capability of the plants. Early infection can reduce yields. Though sulphur sprays can keep it at bay, the best strategy for summer squashes is to have a range of planting/drilling dates, as it affects mature plants more readily. If possible, isolate successive plantings from each other. In wet or humid weather grey mould (*Botrytis cinerea*) can cause rotting-off of young fruit. This can be reduced by increasing airflow around the plants, through wider spacing and/or removing weeds. Cucurbits are affected by a number of viruses, including cucumber mosaic virus and zucchini yellow mosaic virus, which are aphid-borne and capable of stunting and deforming plants. Varieties are available with some resistance to one or more virus diseases.

Harvesting: regular harvesting of summer squash is essential, even if market demand is low, and this may be a daily task at the height of the season, otherwise fruit rapidly becomes unsellable. This requires a considerable labour commitment, even when harvesting rigs are used. Squashes and pumpkins need to be cured in the field or in a polytunnel if they are to be stored (seven to ten days at 25–30°C). They need to be removed from the field prior to any frosts.

Storage: cold storage is useful for summer squash, to remove field heat prior to sale. The longest-term storage of winter squash and pumpkin is achieved at 10°C and 60 per cent relative humidity. The thick-skinned types, such as Hubbard, Queensland Blue and Crown Prince, store best and can last until the next harvest if cured well and stored with care. Make sure that the storage area is free from rodents, who will eat through the flesh to get at the seeds.

Cucumbers (Cuc, *Cucumis sativus*)

Cucumbers are a well-known and productive crop, which is normally grown in greenhouses or polytunnels. A heated greenhouse crop can ensure higher prices. The unheated polytunnel crop can glut at the same time as customers go on holiday and have their own garden crops. All-female varieties are a must. Outdoor varieties can be grown and treated as courgettes and they will crop well in a good summer, but the thick skin, which can be covered in bumps or spines, is often not acceptable to customers used to 'supermarket-type' varieties.

Sowing and Growing: cucumber seed is amongst the most expensive, particularly for the hybrid all-female varieties, and so should be used with care. The cucumber is a semi-tropical vegetable and grows best under conditions of high light, humidity, moisture, temperature, and nutrients. Cultivation methods will vary between cold glasshouse, polytunnel crops and long-season heated glasshouses. For early crops, in heated glasshouses, sow from December onwards at 21–27°C in pots or modules, planting from late December to early February at the eight- to ten-leaf stage. This planting should crop from February/March until September. A mid-season crop, also best in heated glasshouses, can be sown in January for March planting and April cropping. For a late crop, in cold glasshouses or polytunnels, sow in March for April/May planting and harvest from June/July to the first frosts. They need to be planted at least 500mm (20in) apart in the row and can be planted in double rows, with a wider spacing between beds. They need a minimum temperature of 21°C during the day, ventilating at 27°C, for the first two months of cropping. They also generally prefer a higher humidity than tomatoes and are susceptible to cold draughts. Cucumbers are more sensitive than tomatoes to low temperatures, which can cause reductions in both growth and yield. They are indeterminate in habit and can be trained, as cordons, up strings fixed to wires, with the strings buried under the plant at planting time. One way of training them is to take out all side-shoots for the first metre of stem and then stop side-shoots after two leaves. This job needs to be done weekly in summer, as they grow vigorously. Other options are oblique cordons and arch training. Outdoor ridge cucumbers should be sown inside in April for late May/early June planting leading to late July to September cropping.

Soil Fertility and Rotation: cucumbers are highly demanding of nutrients. Sandy loam soils with good levels of organic matter are ideal, as is good aeration and soil structure. Compost or composted manures should be incorporated prior to planting and/or used as a mulch. Commercial, organic liquid-feeds can be used under derogation if necessary.

Irrigation: cucumbers have a high water requirement, as the fruits are mostly water. Watering should be on a little-and-often basis, as growth checks from irregular watering can result in fruits dying back. Drip or trickle irrigation can be used but cold water should be avoided, as it will chill the roots and reduce yields. Watering during the day, when the irrigation lines will absorb heat, will help. It is important to ensure water penetrates to the full depth of the rooting zone. Overhead irrigation can also be used to increase humidity and reduce water loss from the plants, preferably early in the day so that the leaves are dry by night time in order to minimize disease risk.

Weed Control: as for other protected crops, weed control can be through mulching (black plastic or compost), under-sowing or hand-hoeing.

Pests: the main pest is likely to be red spider mite, which can devastate a crop very quickly. Regular monitoring is needed and a biological control, such as the predatory mite *Phytoseiulus*, introduced as soon as possible. Other pests likely to be encountered include aphids and whitefly, for which biological controls are also available.

Diseases: powdery mildew (*Erysiphe* spp. or *Sphaerotheca* spp.) is probably the most important disease. Resistant varieties are the most effective tool, but increased ventilation and removal of any plant debris between crops is also effective in helping manage the disease. Grey mould (*Botrytis cinerea*) can be a problem, especially towards the back end of the season, with cooler temperatures. Where root diseases, such as wilts (*Fusarium* spp.) and corky root-rot (*Pyrenochaeta lycopersici*), are present in the soil, cucumber seedlings of the desired variety and type may be grafted onto resistant rootstocks. The use of compost may also suppress these diseases.

Resistant varieties are available to combat gummosis (*Cladosporium cucmerinum*) and anthracnose leaf spot (*Colletotrichum lagenarium*).

Harvesting and Storage: regular harvesting is needed, every four to five days and the picked fruit needs to be kept cool, preferably with cool (15°C) rather than cold storage. Trays of picked fruit should be covered with plastic in the cool store to reduce water loss.

Melons (Cuc, *Cucumis melo, Citrullus lanatus*)

Melons are a marginal crop that may be useful under protected cropping. The yields are poor, when compared to a crop such as cucumbers. The high unit-value may make them less suitable for a box scheme, but for farm shops and farmers' markets they could be a good crop.

Sowing and Growing: they need high temperatures and humidity throughout the growing period, thriving at around 25°C. The crop is normally sown from March to April, depending on whether a greenhouse or cold crop is required, and planted from April to May, providing the soil and air are sufficiently warm (18–20°C). Plants are spaced at 750–900mm (30–36in) apart and can be trained up strings or allowed to trail on the ground. If trained up strings the leader is tied in to the support and then pinched off, to encourage side shoots that are pinched out a couple of leaves after the female flowers. Support is needed for developing fruit. If plants are allowed to free-range they can be restricted, to avoid congestion or allow access as required. Straw can be put down to prevent contact with the soil. Hand pollination is necessary and five to eight fruits per plant can be expected.

Soil Fertility and Rotation: generally as for cucumbers or other squashes.

Irrigation: regular watering is needed, but avoid the stems becoming too wet. Ridging or raising the beds can help.

Weed Control: control weeds by mulching, under-sowing or hand-hoeing/removal.

Pests: red spider mite, aphids and whitefly may all cause problems, but can all be kept in check with biological controls.

Diseases: as with cucumbers, powdery mildew may be the biggest threat. In humid conditions, gummosis causes sunken areas in the fruit, exuding sticky gum, and causes fruit to crack.

Harvesting: should be from July onwards, when the fruit starts to soften and smell fragrant. Storage potential is limited.

Tomatoes (Sol, *Lycopersicon esculentum*)

Tomatoes enjoy a high year-round demand and so this crop is potentially very lucrative. Outdoor tomatoes rarely do well in our climate and are generally not viable under commercial growing conditions. In contrast, indoor tomatoes can be grown all year round. The simplest growing method is in unheated polytunnels during the summer months when, with careful choice of cultivars and a little additional heat while the seeds germinate, fresh fruit can be available from late June to early November. Days to maturity after transplanting vary from about forty-five for very early varieties to about eighty for some mid-season varieties. In glasshouses the season can be further extended by using heating, although this requires a very high (and potentially environmentally unsustainable) energy input. As with most protected crops, tomatoes are labour intensive to produce and are more prone to pests and diseases than outdoor crops, so they can also be more input intensive. They are a hungry crop, requiring high levels of potassium, and need large amounts of water on a regular basis.

Polytunnel Production: for most growers the unheated polytunnel is the favoured growing situation, therefore is covered in detail in the following descriptions of tomato production. Over-winter and intensive glasshouse production is possible and is briefly described below.

Commercial tomatoes are best grown under protection

Types and Varieties: there are over 3,000 tomato varieties in existence, although those that are commercially available as organic seed number closer to the dozens. They can be divided according to shape and size into beef (or steak), plum, standard and cherry. They also come in different shapes and colours, which add interest to vegetable boxes. Cherries can yield as highly as the larger varieties in terms of weight, but harvest is considerably more labour intensive. The plants can be either bushes or indeterminate types. Indeterminate types are best for yield from a limited planting area (say in polytunnels), whilst bush varieties are more suited to outdoor growing. Trailing varieties are sold in gardeners' catalogues, but are more of a novelty than a commercial proposition.

Seed/Transplant Raising: seed from F1 hybrids is more expensive, although often more reliable. Seed from F1s cannot be saved for subsequent crops, as the plants will not grow true to type, so different priorities will influence the decision on whether to grow them or not. Sowing can start as early as late January or February with heat and the transplants potted on into 90mm (3.5in) pots at the three- to four-leaf stage. The ideal temperature for germination is 25°C, though this should be reduced to 20°C once the shoot is showing. By the time the plants have outgrown their pots, at a height of 150–200mm (6–8in), it may be warm enough to plant them into their final growing position; however, it may be necessary to pot them on again to 130mm (5.5in) pots, if delays with planting are liable to make them pot bound. Tomato plants break easily and must be handled with care.

Grafting: is an option where more vigorous rootstocks are required or in certain disease conditions (such as a cultivar susceptible to verticillium wilt onto a resistant rootstock). It is, however, a fiddly and painstaking task.

Growing: organically, tomatoes have to be grown in the soil and hydroponic or grow-bag systems are not permitted. The key factors to successful indoor tomato cultivation are warmth and ventilation, as well as adequate nutrition and moisture. The plants need a deep bed, so that they can send down the deep roots they need for anchorage, and one rich in organic matter for moisture retention. The best way of achieving this is by the addition of plenty of well-rotted manure. Lime is often added at a rate of 300g/sq m (9oz/sq yd) for ground limestone, although wood ash also has a liming effect and adds potassium as well, at a rate of 0.5–1kg/sq m (14.5–29.5oz/sq yd). Either way, the pH level will need monitoring.

Plants should be spaced at 450mm (18in) in the row to allow for air movement and in single or double rows, depending on requirements and space. For really good ventilation, rows should be 1m (3.25ft) apart. Planting

50–70mm (2–3in) deeper than the soil level in the pot is best, to stimulate new root-growth. If there is a cold snap in the early stages of establishment, the plants may need covering with fleece.

As the plants grow they will require support and there are various ways to provide this. The easiest is a wire running the length of the bed at roof height, which is strong enough to take the weight of the total number of plants in the bed. Strings can be attached to the wire for the tomato plants to grow up. The strings will also need attaching at the bottom end, either directly to the plant or to something structural near the plant base (a cane or second wire, for example). Alternatively, the end can be buried under the plant when transplanting. If the string is to be attached to the plant stem, then it is best to make a loop and thread the top end around the plant stem, then through the loop, before tying it to the overhead wire. This allows plenty of slack for the plant stem to grow into. At the top it is best to use some kind of slipknot, so that the plant can be lowered or the string easily tightened if necessary. If biodegradable twine is used rather than a synthetic material, at the end of the season the whole lot can be added to the compost heap without any fiddly untangling. As the plant grows, string and stem can be wound together and should provide all the support needed. Alternative methods of support include individual stakes or a series of wires running between two posts.

Traditionally it has been the practice to pinch out side shoots. The belief was that this time-consuming activity should be performed regularly, both to increase size and uniformity of fruits and also to allow for increased air movement and manageability of plants. More recent research has implied that if side shoots are left on the total yield remains the same, although the fruit are smaller. Side shoots can also be planted to produce more plants. For ease of management, however, and to prevent the tunnel or glasshouse becoming an impenetrable jungle, it is wise to remove them.

Many growers also stop plants growing once a definite point has been reached, usually either when they reach the roof or after a certain number of trusses have been formed (often six). Others leave the plant to grow, as there is no physiological reason to stop growth at any point, although space may be a limiting factor. In this case, when plants reach an overhead height-limit they can be trained horizontally along supporting wire. Alternatively, as the bottom trusses are harvested the whole plant can be lowered, leaving more growing room at the top. Plants need to be lowered in small increments to avoid snapping the stem and it is best for space management if the stems can be spiralled around on top of themselves at soil level until the lowest fruit is about 150mm (6in) above soil level. The lower leaves need to be stripped off the length to be lowered first and, indeed, this is standard practice to increase ventilation as leaves start to age, as long as there is good top growth. Do not remove leaves above ripening fruit.

In order to ensure a good crop it may be necessary to aid the pollination of the flowers. Tomatoes are naturally pollinated by wind or insects, both of which can be limited undercover. The simple measure of placing flowering plants attractive to pollinators (say French marigolds) at the entrance to the polytunnels, as well as among the tomatoes, may be all you need to achieve good levels of pollination. Mechanical hand-pollinators do exist, alternatively squirting a jet of water into the flowers should aid pollination and it is possible to simply pick an open flower and press it into a dozen or so others to achieve pollination. All of these methods need repeating every time a new batch of flowers opens and consequently they are all very time-consuming. Many commercial growers introduce insect pollinators (often bees) to their polytunnels. The insects are bought in boxes, which are spread at a given density around the cropping area, although these may need replacing every eight to twelve weeks as the insects begin to die out.

Adequate ventilation is very important for indoor tomato cultivation. In the summer, polytunnels need to be open during the day but closed down at night (as the fruit dislike large variations in temperature). As the days start to cool (17°C and below) ventilation can be decreased, although it is still necessary to ensure a through draft on cool damp days to reduce the risk of fungal disease. It is possible that regular (say once per week until fruit sets) spraying of a seaweed extract will help build immunity to pest and disease pressures.

Intensive Glasshouse Production: in a northern, temperate climate, tomato production is only commercially possible under cover, where higher temperatures can be provided. Apart from production in polytunnels over the normal cropping season, it is also possible to produce crops over winter. One of the biggest problems with over-winter glasshouse production is ensuring adequate ventilation and air circulation, as humid conditions among the leaves will lead to fungal and bacterial infections. Spacing may need to be increased and an air gap of 600mm (24in) should be left above the top of the support structure. When plants reach about 1,200mm (4ft), the leaves below the first truss should be stripped off to enable good air-circulation at ground level. The optimum growing temperature for the crop is 25–30°C in the daytime and 15–20°C at night, although fans will almost definitely be needed to ensure that the heat is spread evenly. Glasshouse tomato plants often receive some kind of shading in the summer months, as leaves are easily scorched and fruit suffer several detrimental conditions if they become overheated, although this is uncommon in polytunnel systems.

It is possible to produce two out-of-season crops in one year by sowing one batch in mid- to late-summer, to begin cropping around the time of the first frost, and sowing a second batch in mid- to late-November, to begin cropping in late March or early April. Alternatively, one crop may be produced over winter to take advantage of the price premium. This would be sown in mid-summer and transplanted in September to crop over winter and up to June, when the normal seasonal crop starts. However, it is questionable whether the increase in energy use (and negative environmental consequences) or sacrifice in flavour is worth the outcome. Some varieties have been bred specifically for indoor and/or over-winter production.

Soil Fertility and Rotation: tomatoes are a hungry crop with a long production season, so they need a fertile soil with plenty of organic matter and a pH of around 6.5. Work lots of well-rotted manure or compost into the top few centimetres of soil, repeating several times over the season. Like other fruiting crops, tomatoes have a high potassium requirement, which in an organic system can be supplied by wood ash added to the compost or manure and, later, by comfrey. The latter can be watered on in liquid form or the leaves can be used directly (chop them and allow them to wilt, then fork lightly in or lay on the surface under the plants as a mulch). Organic proprietary feeds are also available and feeding usually starts after the first truss is set. These can be fed through a diluter, but a derogation will be necessary from the certifying body.

Many common glasshouse crops are members of the Solanaceae family, including tomatoes, pepper, chilli, aubergine and okra. It is, therefore, difficult to provide a long rotation gap between them in a glasshouse or polytunnel. A gap of three years is recommended by organic standards, but current standards do allow crops of the same family to be grown successively with prior permission. However, new standards for long-term glasshouse cropping are currently being drawn up.

Irrigation: tomatoes are very thirsty plants, but do not like high humidity, which encourages rots and leaf blight. Therefore, trickle-tape irrigation or similar at soil level is the best form of watering. Grower experience suggests that a large volume of water every couple of days is better than a smaller quantity every day, because it encourages deep rooting. Heavy watering directly at the base of the plant is best avoided, as it washes soil from the shallower roots causing plant stress. In very hot weather the plants will need watering daily or even twice daily. In very hot weather the plant may still wilt, as it will tend to transpire water from the leaves at a faster rate than the roots can take up water from even a moist soil. As long as there is plenty of moisture in the soil, the plants should recover fully when temperatures fall. During hot weather, water early or late in the day, both to reduce water loss from the soil due to evaporation and to reduce the risk of leaf scorch from splashing.

Weed Control: should not be a big problem in the controlled environment of a polytunnel or glasshouse, so that one to two hand-hoeings in the early stage should suffice. After that the canopy should suppress most weeds in the beds and targeted watering should prevent weeds in the pathways. A mulching material, such as well-rotted compost, can be used. Plants can be planted through plastic mulch, although there is a risk of roots overheating under a dark-coloured mulch on hot days if no shading is provided, especially in the early stages of cropping.

Pests: not surprisingly, given the warm and damp environment, pest and disease problems are numerous on tomatoes; from the ubiquitous slug that eats into fruit near the ground through to aphids, eelworms (especially the potato cyst eelworms, *Globodera rostochiensis* and *G. pallida*), leaf miners and various specialized pests, such as the glasshouse whitefly, glasshouse leaf-hoppers and the glasshouse red spider-mite (*Tetranychus urticae*). In the controlled conditions of a glasshouse (and polytunnel) a wide range of sophisticated and effective biological controls exists that can manage many of these pests. Specialist advice should be sought from the relevant companies about running such systems.

Diseases: the biggest potential single threat is probably late blight (*Phytophthora infestans*), which will spread from outdoor potato or tomato crops in humid weather. Tomato blight tends to come later in the year than potato blight and different strains of the disease may be responsible for the disease in the two crop types. Under organic standards it is permissible to spray with copper-based fungicides, provided you obtain prior approval, although the efficacy of the treatment is dubious. Increasing ventilation and avoiding wet leaves under humid conditions will be as effective at stopping the spread of the disease, under most circumstances indoors. Any focus points of disease can also checked by removing and destroying plants (taking care not to spread any spores around). Leaf moulds are fairly common (leaf mould (*Fulvia fulva*), grey mould (*Botrytis*)) and are usually controlled by ensuring the plants are adequately watered and fed. Other common problems include various rots of the foot of the plant (root- and foot-rots), stem and leaf blights and rots (verticillium wilt and sclerotinia disease) and the fruit rots (bacterial soft-rot, blossom-end rot). Some of these are caused by fungi or bacteria, whilst others are more closely linked with nutrient deficiencies. In the case of the soil-borne diseases some resistant cultivars are available, but as long a rotation as possible is also likely to be beneficial.

Viruses can be a problem in tomatoes. They can cause a range of variable symptoms, such as yellowing of leaves, mottling of leaves and curling of leaves. Infected plants are usually weaker and do not produce a good yield, but viruses are hard to diagnose definitively. Viruses are often spread in seed, by aphid, whitefly or nematode vectors, on equipment or by direct contact. The tobacco mosaic virus (TMV), for example, is very widespread and serious. It is spread by contact rather than the usual insect vector and is highly infectious. The virus can be spread from smokers and can survive on clothing for up to three years. It can spread from infected imports so, if tomatoes are packed on a holding for direct sales, particular attention to hygiene is needed and prevention via scrupulous hygiene is usually the best course of action. Some cultivars have resistance to TMV, but should not be grown with susceptible varieties, to decrease the likelihood of resistance breaking down.

Other conditions to be aware of are fruit splitting, usually as a result of irregular watering, ghost spotting, which is caused by splashes of water on the fruit, 'blossom end-rot', which is a symptom of calcium deficiency, irregular ripening, which can be a symptom of potassium deficiency, and 'green shoulder', which is caused by high temperatures.

Harvest: continual picking of ripening fruit stimulates further production and ripening. It also helps to prevent disease, as overripe fruit tends to split, allowing rots to develop. Cold storage will impair flavour, but can be used to remove field-heat from the fruit after picking. It is advisable to leave the calyx intact when picking, because this prevents the invasion of disease through the wound, it also avoids bruising the fruit, it looks better and, as most of the aroma is in the stem, it smells better too. Care must be taken not to tear the stalk or stem: that can allow the invasion of disease. Cradling the fruit in the hand and pressing the calyx firmly with the thumb should ensure that it stays on the fruit. At peak production fruit will need picking daily and is best taken several days before it fully ripens. It will continue to ripen off the plant, whereas a fully ripe harvest will not store.

Storage: the fruit is best stored at 10–15°C under high humidity and it should keep for two weeks or more under these conditions. Some varieties store better than others, so you may need to consider this when choosing seed. Cherry tomatoes do not store well – rarely much more than one week, even in good conditions. Do not layer or stack the tomatoes deeply or the fruit at the bottom are likely to squash and rot. If they are to be stored for longer periods, tomatoes should be checked regularly for rotting fruits and any rotten ones should be removed before the disease spreads.

At the end of the season, large green fruits may ripen indoors – in warmth but out of direct sunlight. On a smaller scale, green fruit can be wrapped in newspaper or something similar and stored in a single layer somewhere cool and dark. They should continue to ripen until about Christmas, although they need regular checking for rots.

POD AND GRAIN VEGETABLES

Pod and grain vegetables often belong to the legume or grass families and can, therefore, be a valuable addition to any rotation as break crops. Legume cash crops – peas and beans – are useful in an organic horticultural system for their ability to fix nitrogen. It is, however, misleading to think of them as fertility builders, much of the nutrients being removed as crop sales. It is better to consider them to be 'neutral' in nitrogen terms, with neither a net gain nor net loss after cropping. They can, therefore, be used to gain an extra season's cash cropping in a rotation.

Runner beans (Leg, *Phaseolus coccineus*)

Runner beans are a popular crop for the home gardener; commercially they are a minor crop but increasingly popular in box schemes. Competition with home gardens is one reason for this, as their peak production period also coincides with the time when many customers are away on annual holidays. A frost-tender perennial, runner beans are generally grown as an annual. They are prolific, but very labour intensive, so for this reason it is best not to grow too many, if unsure of market demand. They are usually grown as a field crop and are not often grown successfully as a protected crop. In the Vale of Evesham, runner beans were often grown in a tunnel due for re-covering, to allow the beans to establish under protection; the plastic being removed as the beans came into flower.

Sowing, Raising and Planting: the key to success with runner beans is to get them established and growing away quickly, but this is not always easy in a cold spring. An early crop can be grown in large modules or pots in late April–May and transplanted. If conditions are cold at planting time they can suffer transplant shock and direct-sown beans can catch them up. They can be covered in fleece to protect from frost and advance the crop but care is needed, as leaves that are touching the fleece can be scorched. They can be sown directly in May or June, as the seed germinates best between 12–30°C. They should be planted in a sheltered location, as they can be damaged by wind.

Growing: if grown up canes, the structures can be vulnerable to being blown over, especially late in the season when fully laden with crop. For this reason many growers choose to train them up individual wigwams of four canes. Double rows should be spaced about 600mm (24in) apart, with poles 300mm (12in) apart and 1–2 plants per pole. At least 900mm (36in) should be allowed between the double rows. If staked as continuous double rows, as in the domestic garden, the row run should not be too long and stout fencing posts should be used for anchorage with wire strained

Runner beans growing up canes in the field

between them. The canes or strings can then be tied into the wires. Poor fruit-set can be due to frosts at flowering time, lack of water, absence of pollinating insects (usually due to poor weather) or sparrows pecking at the flowers.

Soil Fertility and Rotation: runner beans like a light loam, rich in organic matter for water retention, and a good depth of soil. The pH should ideally be around 6.5. If the beans follow a well-manured crop, or after a grass/clover ley, further additions of compost or manure might not be necessary, otherwise 25t/ha (11 tons/acre) of well-rotted manure or compost will be beneficial. One possible strategy is to crop following a grass/clover ley by rotovating strips for the beans to be drilled or planted into.

Irrigation: runner beans have a high demand for water and benefit from regular irrigation. Overhead irrigation will help the flowers to set and irrigation will increase yield once the pods have set.

Weed Control: once the runners have been staked, weed control is restricted to hand weeding. Prior to this, inter-row mechanical cultivations or hand-hoeing are possible. Staking should be delayed until the plants start to run. After this, one-hand weeding will probably be needed. Once established, few weeds will grow between the rows or underneath the wigwam, but the occasional rogueing of the odd fat hen might be needed. Weed control between the double rows can be by cultivations or sowing a legume, such as red clover or trefoil, that is kept mown.

Pests: are not usually a big problem. Establishment can be poor due to the bean-seed fly (*Delia platura*), especially early in the season when conditions are cold and plant growth slow. Black-bean aphid (*Aphis fabae*) can be a problem in some years, but normally predators will keep them under control. If not, then soft potassium soap can be used, sprayed directly onto the insects. In hot, dry seasons red spider-mite (*Tetranychus urticae*) can badly attack the plants, especially if they are not growing strongly. Regular irrigation will help by increasing humidity. A predatory mite can be introduced (*Phytoseilius*), but this will only be effective if conditions remain warm. Birds can also be a nuisance and CDs hung by strings from the canes or other methods can help to keep them at bay, as long as birds do not learn to associate them with food.

Diseases: runner beans are generally free of major disease problems. Grey mould (*Botrytis cinerea*) can develop on the pods when flowering coincides with wet periods and late in the season. Avoid poorly drained and very sheltered sites.

Harvesting: the crop should be ready from July onwards and should be picked at least twice a week. They are best picked when young and tender (pods about 170mm (7in) or so) and not allowed to get too stringy. Harvested produce should be sold as quickly as possible, as it will not store for long periods.

French Beans (Leg, *Phaseolus vulgaris*)

French beans can be divided into dwarf types, normally grown as a field crop, and climbing types, which, although they can be grown successfully outside, are more often seen as a protected or glasshouse crop. They are more delicate than runners and, with smaller pods, are even more labour-intensive to pick. They can, however, give very high returns. Dwarf beans are a relatively quick crop and a late sowing can be fitted in after early crops have finished. A wide range of types and colours are available, with purple and yellow types adding interest for boxes or market stalls. The round ones tend to be most popular, but the flat-podded glasshouse beans, such as Helda, are gaining acceptance. On the marketing side they may have to compete with 'fine beans' from places such as Kenya and Zimbabwe, which are becoming available all year round.

Sowing and Growing: climbing or intermediate types for indoor use can be sown under glass from February (they germinate between 15–30°C) in modules or pots. Avoid blocks, as the seed can rot easily. For planting outside, sow the first crop inside in modules in late April or early May. Drilling can start outside in May if the ground is warm enough. They can be planted in double rows on a bed, with 300mm (12in) between plants in the row and trained up strings to a top wire. They like warm conditions and can be slow to establish. Indoor crops can be harvested from June until the outdoor crop comes on-stream in July and then removed for a green manure or summer/autumn crop. If left in the glasshouse/tunnel a second flush of crop can be obtained by stripping the leaves and liquid feeding (permission will be needed from a certification body for this).

The dwarf types can be sown in succession at fortnightly intervals from May when conditions are warm until early July, aiming at a spacing of 150–250mm (6–10in) within the row and 750mm (30in) between rows. They can also be grown intensively on beds at equidistant spacing, though risk of fungal disease is greater if there is less air movement. They are a relatively quick crop and harvesting can begin in July, relying on the successional sowings for continuity.

Soil Fertility and Rotation: a light loamy soil with good organic matter content is necessary, which will warm up well in spring. As with runners, manure or compost should be added if necessary. The pH should be in the range 6.5–7.5.

Irrigation: is important for yield and fruit set. Care is needed with irrigation to avoid soil splash onto the pods when watering dwarf beans, so as to avoid pod rots.

Weed Control: as for most crops, the better the establishment, the easier the weed control. For climbing beans, treat as runners. For the dwarf beans, stale seedbeds and pre-emergent flaming can be used, followed with mechanical inter-row cultivation. For bed cultivation, they could be planted through a mulch, which would also have the effect of removing risk of contamination with soil splash.

Pests: will be similar to those of runner beans. Slugs may also be a problem for dwarf beans in particular, where beans are close to ground level.

Diseases: similar to runner beans. All beans and peas, but especially French beans, are susceptible to foot- and root-rots, which can be due to a range of soil fungi (*Phytophthora, Pythium, Fusarium*) that attack the plant, normally in cold, wet conditions when the plants are struggling to establish. Rotation will help to keep these diseases from building up in the soil.

Harvesting: should commence outside in July and continue (from successional plantings) until October, if conditions allow. Climbing types must be picked over regularly, every couple of days, to ensure pods don't become stringy and tough. Dwarf beans can be treated the same way, machine-picked or cut and stripped in the pack-house. In larger commercial systems, dwarf beans are often only picked over twice, the second harvest being a destructive harvest to clear the field.

Broad Beans (Leg, *Vicia faba*)

Broad beans are a useful crop as they produce a cash crop early in the season when there is often not much else available. They are hardy and thrive in cool conditions. Sales usually drop off once other green vegetables, such as runner beans, come on tap.

Sowing and Cultivation: if conditions allow, an autumn sowing of a winter-hardy variety (Aquadulce or Claudia, for example) can be done at the beginning of November. They should be deep drilled (70mm (3in)) at 50–100mm (2–4in) apart within the rows and the rows spaced 450–750mm (18–30in) apart. Other varieties, such as the short-podded 'Windsors', can be sown from February. If there is a market then sowing can continue until June.

Soil Fertility and Rotation: broad beans like free-draining fertile soil. Well-worked light land is best for over-winter beans, whereas spring-sown beans can do well on heavier land. Excessive manuring can produce sappy growth, which can lead to frost damage in over-winter crops, so should be avoided. They prefer a neutral or slightly alkaline soil, so liming may be necessary.

Irrigation: is unlikely to be necessary unless exceptionally dry conditions exist as the pods are maturing.

Weed Control: controlling weeds in the autumn-sown crop can be difficult, as conditions usually don't allow for mechanical weeding or ridging. Hand hoeing may be necessary. For the spring-sown crop a comb harrow can be very effective, just don't look back!

Pests: the over-winter beans tend to avoid the worst aphids (generally black bean aphid (*Aphis fabae*)) but spring-sown beans can be susceptible. Eventually the ladybirds and other predators will take over and clean up. Pea and bean weevils (*Sitona lineatus*) can nibble the edges of leaves, to give a serrated effect, but damage is rarely serious. Mice can be a nuisance, gnawing holes in the pods.

Diseases: the most severe disease problem is likely to be chocolate spot (*Botrytis fabae*), which is worst on over-winter crops and in wet seasons. Adopting wider row spacings can help and avoiding poorly drained, very sheltered or very exposed sites. If possible, separate winter and spring sowings geographically, to avoid transfer of the disease, and remove any volunteer bean plants from previous years. Rust (*Uromyces fabae*) is common late in the season and can develop rapidly in warm weather, but has little effect on yield.

Harvesting: avoid the mistake of letting the pods over-mature before picking as the young beans are the tastiest. They are best sold quickly.

Peas (Leg, *Pisum Sativum*)

Peas for fresh picking are a labour-intensive crop, but can be worthwhile if the market is available. They are suitable for cooler climes and are useful as an early crop, when few other vegetables are available. Mangetout and sugar-snap varieties can command a good premium and are popular for direct marketing and for the restaurant trade. Some of the taller sugar-pea varieties, such as Sugar Snap and Oregon Sugar Pod (Mangetout), can be grown up netting in polytunnels for an early crop. Vining peas are usually grown under contract for the frozen-food processing companies, close to the factories.

Sowing: the earliest sowings can be made in November, in favourable locations, using round-seeded varieties (such as Feltham First) or, alternatively, in very early spring, as conditions allow. The wrinkle-seeded types can be sown from March until June. A good tilth and firm seedbed is necessary. For shorter varieties use a narrower spacing of 250mm (10in), which will allow the plants to support each other. A wider spacing of 600–1,200mm (2–4ft) apart is needed for the taller types. The seeds should be drilled at 10–20mm (0.5–0.75in) apart in the rows.

Growing: care needs to be taken with choice of variety, as the taller types will need staking and nets as support and this is labour intensive.

Soil Fertility and Rotation: peas favour a well-drained soil with good potash levels. Manuring should not be necessary. The pH should be 6–6.8.

Irrigation: may be necessary in dry seasons.

Weed Control: can be difficult to manage in peas. If a weed strike is possible prior to drilling, this will help. A comb-harrow can be used once the peas have emerged and are about 60mm (2.5in) high. Inter-row weeding can be done in the early stages and ridging for the taller varieties prior to staking.

Pests: mice can be a problem, eating the seeds after sowing, and precautions will need to be taken against birds, either by using scarers or by covering the sowings with fleece or netting. Birds will also attack the pods and remove seeds. Pea moth (*Cydia nigricana*) larvae can also be a problem, as they burrow into the pod and feed on the peas. Control is difficult, but good rotations should keep levels down and exposing crop debris to weather and predators will help reduce over-wintering caterpillars. Crop covers can be used, but only work if in place prior to the infestation. Early and late crops are less prone to infestation, as the main egg-laying period for the moth is June and July.

Diseases: peas seem prone to a number of diseases. Downy mildew (*Peronospora viciae*) is a common soil-borne disease that can infect seedlings and stunt growth but should be controlled by a long rotation (more than five years). Resistant varieties are available. A range of soil-borne fungal diseases that cause foot rots can infect peas, usually associated with intensive legume cropping and poor soil structure. A number of diseases can cause small spots on the pods, leaves and stems and reduce crop quality. Healthy seed is important. Grey mould (*Botrytis*) can infect pods when wet weather occurs during flowering and petals stick to the pods.

Good air movement around the plants, through wider row spacing can help alleviate problems. Powdery mildew (*Erysiphe polygoni*) can colonize plants in dry seasons but might not have a large effect on yield, though appearance for marketing can be a problem. Good hygiene to destroy residues and isolating successional sowings from each other can help.

Harvesting: the dwarf crop is normally cleared at one picking, whereas taller varieties may be picked over a number of times. The over-winter crop should be ready to harvest in June, following into early spring, and maincrop varieties until September, if required. Freshness is the key and direct marketing has the advantage. Mangetout peas should be picked as soon as a convenient size, when the peas can be felt through the pod. Picking should continue regularly to encourage production.

Sweetcorn (O, *Zea mays*)

Sweetcorn can be successfully grown in the UK and is a popular crop with consumers. It is best suited to the growing conditions of the south and east of England, but can be grown elsewhere by using earlier maturing varieties and by transplanting. As with all crops, you need to be sure of your market. It is particularly useful for direct marketing, although the season is relatively short. It has the advantage of not being in any crop group normally grown in the rotation.

Varieties: at present, the range of varieties available as organic seed is limited and includes few that could be described as commercial. Sweetcorn varieties can be classified as normal through to super sweet. The normal, sugary types have the most rapid conversion of sugars to starch after picking and, consequently, a short shelf-life. They are early and germinate better in cold soils, but don't normally yield as well as sweeter types. The super-sweet varieties are, as their name would suggest, very sweet! They have very slow conversion of sugars to starch and, thus, the longest shelf-life, but germinate poorly in cold soils, so are usually best raised as transplants and planted out. The new generation of Xtra-tender varieties are said to be even sweeter than super sweets, with a tender pericarp and excellent eating qualities. It is important to isolate the super sweet and Xtra-tender types from the others (80–250m (250–800ft) is needed), as cross-pollination can result in starchy kernels.

Sowing and Raising Transplants: sweetcorn germinates best between 15–35°C. Therefore early crops, super sweet and Xtra-tender types are best raised as transplants, for more reliable establishment. In more northern areas, all sweetcorn will need to be raised as transplants. Sow the earliest varieties in modules or blocks from early April for transplanting at

Sweetcorn is popular with direct-sales customers

the beginning of May under fleece. To some extent, succession can be managed by varietal choice, but sowing can continue until May. It is important that transplants are not allowed to get too big (beyond the two-leaf stage), as they can suffer greatly from transplant check. Direct-drilling is also possible, as conditions allow, until late May.

Growing: sweetcorn needs a long season and warm temperatures to do well. Spacing should be around 250–750mm (10–30in). One system used in the USA is to cultivate strips into a grass/clover ley and sow or plant into that. This could be done with a pedestrian rotovator or by removing rotor blades on a tractor rotovator. To prevent excessive re-growth, the clover is 'partially rotovated' two weeks after corn emergence. Intercropping is also possible with pumpkins, squashes or dwarf beans, but attention is needed to ensure plants don't climb up the stems and obscure the flowers, interfering with pollination.

Soil Fertility and Rotation: sweetcorn is a demanding crop, especially of nitrogen, and therefore it fits best early in the rotation, following a grass/clover ley or after an over-winter vetch crop. It does best in a deep well-drained soil with a pH of 6–6.5 and needs moderate levels of phosphorus and potassium.

Irrigation: sweetcorn is relatively drought-tolerant but yields will be increased by irrigation during ear development.

Weed Control: stale seedbeds and weed strikes or flaming, prior to emergence or planting, will give a head start over the weeds. Thereafter the options are mechanical inter-row cultivations and under-sowing or a combination. Cultivations must be shallow to avoid root damage. Good weed control is important, as weeds can interfere with crop pollination, and thus crop quality, if allowed to get away. Under-sowing with a low-growing legume, such as white clover or trefoil, is possible, usually after the first weeding-operation. A later sowing of red clover is another option. Some trials have been done that suggest that post-emergence flame-weeding is possible, flaming at a crop height of 100mm (4in) for short-term weed control.

Pests: birds, particularly rooks can be damaging and, if they are a problem, scarers or crop covers will be needed before birds become accustomed to the food source. Leatherjackets (moth larvae of *Agrotis* spp.) can be a problem after grass leys and they can be reduced by cultivations and exposure to the birds (that will later eat your crop!). Aphids can be a problem, making the cobs unsellable, but are usually dealt with by the beneficial insects if left alone. Once the cobs are ripening, rodents and badgers are the biggest risk. An electric rabbit fence may need to be erected against the latter.

Diseases: there should be few diseases to worry about with sweetcorn, provided good organic rotations are practised. Smut (*Ustilago maydis*) on the cobs can be spectacular but is seldom a huge problem. Stalk rot (*Fusarium* spp.) can cause wilting and premature yellowing from the flowering stage.

Harvesting and Storage: trimming is normally carried out in the field. Picking in the morning and cold storage is advisable, or at least rapid removal of the field heat from the crop. On a larger scale, tractor-mounted packing rigs can improve harvesting efficiency. Fresh cobs should be sold quickly, as they do not store for long periods.

BULB AND STEM VEGETABLES

These vegetables mainly comprise alliums and umbelliferas and, as such, are often placed later on in the rotation; in many cases they are fairly winter-hardy. The alliums are a major cropping group, including onions – a staple crop – and leeks, which are a mainstay of winter growing programmes. They all have shallow root systems (300mm (12in) or less), sparse canopies and are tolerant of frost. Most alliums are also very sensitive to day length. The umbelliferas are represented by celeriac and celery, as part of the plant family that also include the root crops carrot and parsnip. The bulb vegetables can store well and, as a group, the bulb or stem vegetables form the basis for the flavour in many recipes.

Celeriac (Um, *Apium graveolens rapaceum*)

Celeriac is the swollen or 'turnip-rooted' ugly sister of celery. There is a limited market for it but it has the advantage for growers that it is hardier and more robust than celery and, with storage, can be sold through the winter.

Seed and Raising Transplants: the seed is small, irregular in shape and difficult to sow. Celeriac should be sown into blocks or modules from February (with heat to create warm conditions).

Planting and Growing: celeriac needs a long season and should be planted out from May to mid-June, 300–400mm (12–16in) apart, depending on market requirements. Care is needed when planting to plant them as shallowly as possible, to avoid burying the crown, or they can become a mass of roots that will need trimming off. Most enlargement of the root occurs from September onwards.

Soil Fertility and Rotation: as for celery. It can be sensitive to low boron-levels, especially on sandy soils.

Irrigation: as with celery it needs to be kept growing and does not like to dry out. It will, therefore, benefit from regular irrigation.

Weed Control: as for celery.

Pests: subject to similar pest problems as celery, though slugs are likely to be less damaging (to the leaves) and carrot fly more problematic, as the root is sold.

Diseases: celeriac can also suffer from leaf spot (*Septoria apiicola*), but it is not usually such a big problem, possibly due to a more open growth-habit than celery.

Harvesting and Storage: an early celeriac crop can be harvested from August, but if celery is also grown it can be left until late September or October, so as not to clash. Celeriac can take some frost but needs to be protected by straw for fresh digging in the field from November or lifted and clamped.

Celery (Um, *Apium graveolens*)

Given good conditions and the right season, celery can be a very successful organic crop. An early crop can be grown under glass or in a polytunnel. In the past, hardy trenching celery was widely grown on the fens for autumn/winter harvest, but now the less hardy self-blanching types are grown. Whereas twenty years ago, or less, self-blanching white celery was the fashion, the trend is now almost universally for green types (such as Victoria or Granada).

Seed: the biggest issue with celery, currently, is with regard to seed quality. Celery leaf spot (*Septoria apiicola*) is seed-borne and, until recently, many organic growers obtained derogations to use treated conventional seed, which is no longer permitted. It was considered that organic celery would not be able to be grown without this derogation. Organic heat-treated seed is now available, though germination can be erratic with current treatment methods.

Raising Transplants: celery is normally raised as blocks or modules and needs a long growing season. It is best not to cover the seed, as light is required for germination. It can be sown from early February in the glasshouse until April, needing 10–22°C for germination. Successional sowings should be made at ten to fourteen day intervals from early March for outdoor plantings. Good hygiene is needed, as the seedlings can be susceptible to damping-off diseases soon after emergence.

Planting and Growing: the soil needs to be deeply worked for celery. A protected crop could be planted in April and the outdoor crop from early May (under fleece) to early July. Spacing should be around 250–300mm (10–12in). It is important to keep celery growing steadily, as it will not tolerate any checks to its growth. Some growers double-crop celery, planting a late crop after the early one has been harvested but this is not recommended, as leaf spot is likely to carry over between crops.

Soil Fertility and Rotation: a fertile, deep soil, rich in organic matter, is best for celery, though it will grow in light sandy soils given adequate manure and plentiful irrigation. It will not tolerate compaction or poor soil structure. The ideal pH is between 6.5–7.5.

Irrigation: celery is 94 per cent water and has developed from a marshland plant. Not surprisingly therefore, it has a high requirement for water and needs regular and uniform irrigation.

Weed Control: celery can be slow to establish and weed control needs to be good in the early stages, after which celery covers the ground and effectively out-competes weeds. Usual inter-row weeding operations, such as brush-weeding or steerage hoeing, can be carried out.

Pests: celery leaf miner (*Euleia heraclei*), the larvae of the celery fly is likely to be encountered; it burrows into the leaves but the damage is not usually severe and can be easily trimmed off. Leatherjackets can be a problem after grass/clover leys, as can slugs. Effective mechanical weeding can keep slugs down. Celery can also get carrot fly (*Psilia rosae*), but this insect is not generally considered a problem in this crop.

Diseases: as mentioned above, the most serious disease problem is celery leaf spot (*Septoria apiicola*). Severe attacks can cause browning of the foliage and disease progression can be very rapid in wet and windy conditions, leading to celery 'blight.' If present on the petioles it can render the crop unmarketable and it can develop in the humid conditions of the plastic celery-sleeve, post-harvest. Healthy seed and transplants are essential. Once leaf spot arrives, the crop needs to be moved as quickly as possible. Copper sprays are presently permitted under derogation but are only useful if applied at a very early stage. Successional plantings should be isolated from each other to reduce spread. Celery can also be very susceptible to pink rot (*Sclerotinia sclerotiorum*), which is a soil-borne disease that develops following an initial water-soaked rot of the leaf stalks. Rogue out and burn any infected plants. A number of viruses can affect celery, including celery mosaic virus, but they are not normally too serious.

Harvesting and Storage: the early celery plantings under protection can be harvested from June. The outdoor plantings can be harvested from July to October, until the point that they succumb to leaf spot or become frosted. A good, sharp knife should be used and if you can find a specialist celery-cutting knife (as used on the Fens) then this is ideal. Being mostly water they can lose moisture and deteriorate quickly after harvest, unless put into plastic sleeves and cold stored to remove field heat. Once ready, they will not stand long in the field and need to be monitored to avoid selling stringy or pithy sticks.

Onions (All, *Allium cepa*)

Onions are a major crop but not an easy one to grow in the UK, especially as they have to compete with cheap foreign imports. They are a field-scale crop and by using a combination of varieties and storage they can be made available all year round. Over-wintering onions can be grown for an early crop in late June and July. Maincrop onions are initially produced from sets and later from modules or by direct sowing. With good storage, onions can keep through to May or June of the following year. Red onions are becoming more popular with consumers looking for more eye-catching onions in

Harvesting and packaging celery in the field

salads and fuelled by TV cookery programmes. Onions are sensitive to day length, which governs bulb initiation. They need to complete the majority of their vegetative top growth before the longest day, which triggers bulb formation as the days start to shorten.

Sowing and Raising: the choice for onions is between direct drilling, which is the method of growing onions conventionally, multi-sown modules/blocks and sets.

Direct drilling: is possible on light soils, providing they have low weed pressure, that can be cultivated to a good tilth for a February or March sowing. The cost of establishment is lowest for this method but good weed control is critical and it is risky. The seed germinates above 5°C, but, in a cold spring, they can be slow to germinate (from three to six weeks depending on weather conditions) and to grow away. In this period they can be vulnerable to weed competition. Seed priming can have a positive effect on speed of emergence and on crop yield. Seed should be sown at about 40 seeds per metre or 25mm (1in) apart. Row spacing (400–600mm (16–24in)) depends on the bed system being used and the weeding equipment available, as well as the soil fertility, to some extent.

Modules/blocks: onions can be raised by sowing five or six seeds in a block or module in February or March under glass for an April planting. Plants raised from seed are thought to provide better disease resistance to downy mildew (*Peronospora destructor*), which is a particular problem, and to therefore show better storage quality. They can be sown early, independent of soil conditions, but need care and may need extra feeding, for which a derogation will be required, unless a large size of module or block is chosen. There is an advantage for weed control in that they are planted later (but no later than April) and weed strikes can be carried out prior to planting. They are also planted further apart than sets, enabling hoeing within the rows. A range of sizes and shapes can result from using transplants and it is important to have no more than four or five plants per module or block. They should be planted out at 200–250mm (8–10in) within the row and 400–600mm (16–24in) between rows.
Sets: there are now a number of different varieties available as organic sets, which are part-grown bulbs. Planted in March or April, sets are quick to establish and most competitive against weeds. Plant spacing will determine the size of bulb, 50–80mm (2–3in) apart for small bulbs, 150mm (6in) apart for large ones. They can be prone to bolt, if conditions are not to their liking. Onions grown from sets usually mature at least two weeks earlier than sown onions, around the middle of August.

Soil Fertility and Rotation: onions can be grown well on a wide range of soil types. Light, sandy soils are ideal however, as they warm up quickly for early growth and maturity and bulb-staining is reduced. Onions require a soil pH between 6.2–7.5 on mineral soils, or 5.4–7.5 on peat. Onions ripen later and more slowly on soils with high organic matter content. Stones can cause problems at drilling and at harvest. Good drainage is essential, but onions need moisture throughout the growing period and soils must have good water-holding capacity. Soils with high levels of calcium are shown to reduce the risk of black mould (*Aspergillus niger*) in storage.

Irrigation: is useful for onion production, particularly when grown on light soils, and can be essential for establishment of transplants. Irrigation can decrease water and heat stress, making onions more uniform and less vulnerable to pests (such as thrips) and diseases. Overhead irrigation may increase risk of bacterial diseases, however, and uneven application of irrigation water can cause erratic emergence, growth and ripening. It is best not to irrigate within two to three weeks of harvest, as this can induce disease in the neck and delay maturity.

Weed Control: onions are very sensitive to weed competition during the vegetative-growth stage. The direct-drilled crop relies on effective flame-weeding for success. The preparation of a false (stale) seedbed is unlikely

Onions need to be weed free during the early crop stages

to be practical in the spring sown crop, but the period from drilling to crop emergence is usually long enough to provide an opportunity to kill early emerging weeds before the crop emerges. If conditions allow, delayed drilling of the crop gives weeds more time to emerge and to be killed by flame weeding before crop emergence. Inter-row weeding needs to start as soon as the rows are visible, using a brush weeder or steerage hoe. The critical period for weed competition appears to be around the third true-leaf stage, five to eight weeks after crop emergence.

Cultivations or weed strikes can be made prior to planting modules or sets, or a stale seedbed can be created and flame-weeding employed. Flaming can also be used post-planting or -emergence using shields or directional burners. Inter-row cultivations with brush weeders, tine weeders, harrows or tractor-drawn hoes will control weeds between the crop rows. Tine weeders and torsion weeders can also help to work the soil within the row, reducing hand-labour requirements.

It is usual for onions to require hand weeding at some stage. Module-sown onions can be hand-hoed within the row. The optimum timing of hand-weeding for module-planted onions is around five to seven weeks after transplanting.

Pests: the principal pests are likely to be onion fly maggots, stem nematodes, cutworms, wireworms, thrips, rabbits and birds. Rabbits and birds are particularly troublesome when they pull sets or transplants out of the ground, causing poor establishment.

The stem nematode (*Ditylenchus dipsaci*) has a wide host-range and affects alliums by causing twisting, distortion and discoloration of stems and foliage, and distorted and cracked bulbs. Infection can be introduced by contaminated seed or planting material, so it is vital to obtain clean seed. Infected onion sets tend to be soft, shrunken and discoloured (dark brown) near the neck of the plant and are lighter in weight. Severe infestations tend to be worst in wet-weather periods. Rotation is important, as nematodes can survive in moist soil for about a year. The use of compost can have a suppressing effect on nematodes.

The maggots of onion fly (*Delia antiqua*) can cause problems, particularly in the eastern counties. They over winter in the pupal stage in previous allium fields and emerge in May. The flies are responsive to onion odours, lay their eggs on plants or soil and the maggots emerge to feed on the bulbs or leaves. Attacked leaves become yellow and wilt, and seedlings may disappear. Good sanitation, post-harvest ploughing, crop rotation and siting crops distant from past infestations will help. Crop covers could also be used, during the most vulnerable periods.

Diseases: the main disease problems likely to be encountered on onions are white rot, downy mildew and neck rot.

White rot (*Sclerotium cepivorum*) affects all alliums and is the most damaging disease. The first signs are leaves yellowing and wilting. When the plants are pulled up they will have a fluffy white rot, starting from the base. Onions are usually affected more seriously than leeks or garlic. It is a very persistent soil-borne fungus and can remain viable in the soil for up to twenty years. It can be spread by cultivations and any infected plants should be removed and burnt if practicable. There is some evidence that composted onion-waste can help to manage this disease, by stimulating organisms that attack the pathogen.

Downy mildew (*Peronospora destructor*) is principally a disease of onions, manifesting itself as purplish-grey spores on large leaf-blotches, which may be at the leaf tip. Given the right conditions (warm and humid), it can spread very rapidly through the crop, causing collapse of foliage and secondary, sooty moulds. If infection occurs early, which is more likely after a mild winter, it can seriously reduce yields and result in soft, immature bulbs. Sheltered sites should be avoided and successive plantings should be isolated from each other.

Neck rot (*Botrytis allii*) is a disease that occurs in the field and continues into storage. It is a seed-borne fungus, but could also be carried in sets of onions and shallots. Although it can be present at harvest, it is more often observed in storage, when a soft, brown rot develops from the neck. Topping of the foliage prior to harvest can allow an entry point for the disease. The main control-method is to prevent the fungus growing down the neck of the bulb after topping. Topping to leave a neck of about 80mm (3in) and rapid curing is advised.

Botrytis leaf spot and leaf rot (*Botrytis quamosa* and *Botrytis cinerea*) cause small white spots on the leaves, with a water-soaked margin or halo, during wet or humid periods. It is particularly a problem with salad onions, where the blemishes can make them unmarketable.

Harvesting: onions are ready to harvest when the tops begin to fall over. After 15–25 per cent of the tops have fallen over they can be harvested for immediate sales and when 50 per cent have fallen they can be harvested for storage. They need to be properly cured in order to store. The traditional system of harvesting is probably best for organic growers; it involves undercutting and windrowing until the tops have wilted (five to ten days). Good weather at this juncture will be beneficial. Complete curing in the field is unlikely to be achieved in the UK's fickle climate, as temperatures of around 25°C and 60–70 per cent humidity are needed for two to three weeks. An eye needs to be kept on the weather forecast, as wet weather during this process can cause skin discoloration and loss of quality. Maincrop onions should be off the field before mid-September, to ensure best quality, and into store for drying. Alternatively, the standing onions may be topped in the field (to 80mm (3in)) and harvested directly for drying instore. Sprouting of bulbs instore is encouraged by this early defoliation. Care must be taken during harvest to avoid bruising.

Storage: on a small scale, after a degree of field drying and curing, onions can be laid out on trays or pallets in a polytunnel to complete the process (earlier if bad weather). Artificial drying and curing needs temperatures of 28–30°C, with constant air-movement for two to four days, to promote strong outer skins and neck closure. This should be followed by a second drying period of about two weeks at 26°C and 70–75 per cent humidity. After this, the crop should be ventilated with ambient air to avoid condensation. A low-tech and low-cost solution is to lay onions over a mesh frame and blow warm air through with an industrial space-heater.

Another approach is for trays or boxes to be stacked to form two walls either side of a drier or fan to force air through.

For longer-term storage, careful selection of varieties is important, choosing long-dormancy characteristics and a good outer-skin covering. Smaller onions also tend to store better than larger ones. Ambient air storage is sufficient until December, or possibly as late as March with the right variety, in ventilated clamps or racks in barns. After that, cold storage is necessary, at 0–7°C and 70–75 per cent humidity. Care has to be taken to avoid freezing damage, which can occur at –1°C, and stores should be inspected regularly. When the onions are removed from store they should be conditioned for several days at 20°C and 50 per cent relative humidity.

Specialist Onions: various specialist onion crops exist and can be developed to exploit specific market opportunities. They include producing over-wintering onions and shallots or spring onion production for fresh consumption.

Over-wintering onions: sometimes referred to as Japanese onions, these have a useful function in providing fresh, early season onions in late-June and July, but they are not the easiest crop to grow. They are best grown as multi-sown modules, sown in the second half of August for a September planting, or from sets planted in September/October. They are likely to suffer a heavy weed burden from both autumn and spring flushes of weed emergence and mechanical weeding can be difficult over the winter period, so they can benefit from planting through a black-plastic mulch. Spacing through mulch can be an equidistant 200–250mm (8–10in) for modules. Sets should be planted at 110mm (4.5in) apart in staggered rows 150mm (6in) apart. It is important that the plastic is laid tightly to avoid flapping or movement. The other issue with over-wintering onions is supplying nitrogen for early growth in the spring, when the soil biological activity is low. The plastic mulch will help warm up the soil, but a derogation to apply a supplementary source of organic nitrogen may be needed, if they are suffering unduly. The mulch can help with field wilting, due to the plastic absorbing heat and the absence of weeds. The over-wintering crop will not store for long and so is best used to provide fresh onions for sale before the set-planted crop comes in.

Shallots (Al., Allium cepa*)*: sweeter and spicier than onions, a small demand from foodies exists for shallots. They are multi-centred, having several growing points, but can be treated as onions, culturally. Longer day varieties are available that are suitable for growing in the UK, with good storage properties. Plant sets as early as possible, as soon as the soil has warmed up, from the end of February. The aim is to get uniform shallots (between 30–45mm (1.25–1.75in) for supermarket buyers). Space about

100mm (4in) apart, with rows 450–600mm (18–24in) apart, according to what best fits in with your system. They should be planted 25mm (1in) deep and, preferably, with the basal plate downwards. They can be harvested and stored as onions.

*Salad onions (*Allium cepa *and* A. fistulosum*)*: also a difficult crop to succeed with, due to difficult weed control and maintaining continuity and quality. The trimming and bunching process can be time-consuming and costly, especially if the quality is not top-notch. Salad or spring onions are quick growing, however, and can provide an early crop in the spring, when fresh produce is scarce. They are harvested in the immature stage. Usually it is cultivars of *Allium cepa* that are grown as salad onions, however, Japanese bunching varieties of *A. fistulosum* are sometimes grown, but may have limited winter hardiness. The Japanese varieties and *A. cepa* × *fistulosum* hybrids, such as Guardsman, are less susceptible than White Lisbon types to downy mildew (*Peronospora destructor*), which can obviously be important where the produce is eaten fresh. Winter-hardy varieties can be sown in modules (6 or 7 seeds per module) from July to August for August to September planting, 150mm (6in) apart in the row, in the field or in polytunnels. The spring crop can be sown in modules from February and direct drilling can be carried out from March through to June, at 65–70 seeds per metre row length (20–22 seeds per foot), with the same provisos as for bulb onions. Sowing or planting should be at two- to three-week intervals to ensure continuity. Considerable hand labour (1,000 hours/ha) is required for harvesting the crop. At harvest, salad onions are usually bunched in the field, then taken and washed and trimmed elsewhere, before being packed in containers.

Leeks (All, *Allium porrum*)

Leeks are a popular vegetable, which are relatively easy to grow and can be harvested from late June/July through to May the following year, from a range of planting dates and varieties. This makes them a very useful crop for direct sales, though this can sometimes be regretted when harvesting in the freezing rain in the depths of winter.

Seed/Transplant Raising: it is important to choose the correct variety for the appropriate season. The choice for plant raising is between modules, bare-root transplants and direct drilling.

Direct drilling: as with onions, direct drilling has not been a popular way to grow organic leeks, due to the uncompetitive nature of the crop and its slowness to establish. The high cost of transplants and planting has led some growers to try it, but it is risky and, if flame-weeding and subsequent mechanical weeding are not carried out at the right time, it can be costly for hand-weeding to control the subsequent weeds. Seed-priming can be used for faster emergence. Drilling can take place from mid-March to May, aiming for spacing of between 40–80mm (1.5–3in) apart in rows with 300–500mm (12–20in) between plants. Wider spacing is better where ridging for weed control is intended.

Bare-root transplants: there are advocates of bare-root transplants for organic production, whereby the leeks are sown at a high density in a seedbed (90–120 seeds per metre (28–36 seeds per foot)) and transplanted at pencil thickness. It is best only to do this on the holding, as bought-in bare-root transplants can carry risk of transferring disease. They can be labour intensive for weed control in the early stages, and stale seedbeds and pre-emergent flaming should be used. They do allow more flexibility in planting time than modules, if conditions are unsuitable for planting, as they will grow bigger in the soil, rather than stagnating in a module. An early sowing can be made in a glasshouse or polytunnel. Leaves should be trimmed to reduce transpiration, reduce transplant shock and prevent the leaves trailing on the ground and becoming a source of infection.

Modules: leeks are not the easiest plant to grow as an organic modular transplant. This can be important, as the quality of the plant has a large influence on the quality of the end crop. Plant raisers are, however, getting better at raising leeks. Multi-sown modules or blocks were in vogue a few years ago, aiming for two to three plants each. This may still be suitable for box schemes and direct marketing, where a slight unevenness of shape and colour might be acceptable, but less so for supermarkets. Single-sown modules are now widely used. As with onions, the transplants may sometimes need supplementary feeding in the modules.

Growing: the earliest leek plantings can be made from April, into a well-cultivated seedbed. Planting can continue with maincrop leeks planted in June and early July. Modules can be mechanically planted or dibbing machines can be used that make the holes, followed by hand planting. The highest quality will be achieved by planting into dibbed holes and this is essential for bare-root transplants. The soil needs to be moist for the dibbed holes to stay open, but not so wet that compaction is caused by planting or that the holes 'smear'. On a smaller scale, the holes can be hand-dibbed or a frame adapted to dib a number of holes at once. The holes can be up to 150mm (6in) deep and 50mm (2in) across and the plants are simply dropped into the hole and not firmed. The holes will fill with soil from cultivations and irrigation. They can be planted at 80–150mm (3–6in) apart, in the row, depending on market requirements, and watered-in as soon as practicable. The row spacing depends on weeding regime. They are often grown on a bed system, to facilitate harvesting, using an under-cutting bar.

Soil Fertility and Rotation: leeks can be grown on many soil types but deep, fertile loams are best for producing good yields of high-quality produce. Good drainage is essential and the soil should not be compacted. The pH should be above 6.5. Leeks are not a demanding crop, but will respond well to generous applications of bulky organic manures. Leeks can follow a cereal or be the second vegetable crop in the rotation. The maincrop is planted late enough to follow an early crop, such as broad beans or an extended period of winter vetch.

Irrigation: irrigation will be necessary for planting and establishment and to prevent checks to growth, which can lead to bolting. It can also reduce damage by thrips and cutworms.

Weed Control: stale seedbeds (if time permits) and pre-emergence flaming are necessary for the drilled crop. For the transplanted crop, more time generally permits a number of weed strikes prior to planting, which should help reduce weed pressure. Inter-row cultivations with brush weeders, tine weeders, harrows or tractor-drawn hoes can be used to control weeds within the crop row. The number of passes will depend on the weed population. With many of these implements the soil may be directed towards the crop row to bury small intra-row weeds, as well as improving stem blanching. Care should be taken not to get loose soil down within the crop leaves. Where inter-row cultivation is used, leek rows are spaced, typically, at 400mm (16in) apart. At wider row spacing, a potato ridger may be used to earth-up the crop and aid weed control. As with onions, post-emergence flaming using shields or directed burners is possible.

Pests: thrips (*Thrips tabaci*) are commonly observed on leeks, normally due to their damage, which is characterized by white or silvery flecks on the leaves that cause the plant to look unsightly. Leaves may later become distorted. Thrips overwinter in the soil and ploughing in crop debris can help, as can separating allium crops by space and time (rotation). They thrive in hot, dry conditions and irrigation can help reduce symptoms. Damage is usually worse on leeks, where the silvery flecking on the green parts of the leaves can spoil cosmetic appearance, reduce marketable yield and necessitate extra trimming. There is some research that suggests coloured mulches may be effective in their control and that might be worth a try if mulching anyway for weed control. Leek moths (*Acrolepiopsis assectella*) occasionally cause serious problems, particularly in hot, dry weather and in the south of the UK, by mining the leaves of the leek and causing unsightly damage, rendering the plant unmarketable. Management methods mainly rely on cultural methods, such as destroying crop debris or removing foci of infection. Cutworms, wireworms and rabbits may also cause problems.

Diseases: rust and white tip are the principal disease problems observed in leeks, although the other onion diseases do also occur (*see* above). Rust (*Puccinia allii*) is seen as small, bright-orange pustules that appear from July onwards in warm and wet conditions, reducing marketable quality and causing death of foliage when severe. Degrees of varietal resistance exist and careful choice is needed. Isolating successive plantings can help to reduce risk of spread. White tip (*Phytophthora porri*) damage consists of water-soaked white lesions at the tips of leaves and dieback in late summer or autumn, particularly of leeks. It is a soil-borne disease that can survive for several years. Rotations should be extended if it is a problem.

Harvesting and Storage: is a time-consuming, labour-intensive and often uncomfortable task. The earliest plantings can be harvested from late June/ early July, if there is a market. The main harvest period is from September through to April or May, when the crop will start to bolt. Leeks can be harvested by hand with a fork, or by using an undercutting bar to loosen them from the soil prior to lifting and trimming by hand. Care needs to be taken, as damage can be done to soil structure if heavy machinery is used in wet soil conditions. Leeks are normally trimmed in the field, though if weather is inclement they can be transported to a barn for trimming. They can be stored in this state, prior to trimming, for two to three weeks, which can be useful if a big freeze-up is forecast. However, it is preferable to

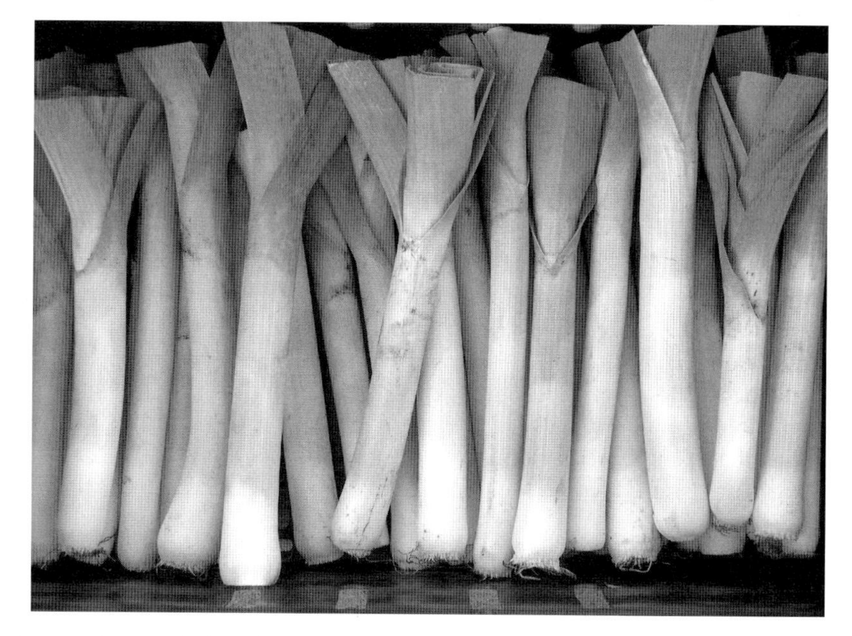

Trimmed leeks ready for sale

prepare them as near to their marketing state as possible. It is usual to trim the roots off to leave the hard pad at the base, peel one or two outer leaves from the shank and to make a top cut above the shank according to market requirements. For direct marketing, provided the leaves are disease free, more green can be left on the plant. A sharp knife is essential, to get a clean cut – keep a sharpener handy, as cutting the roots soon dulls a blade. It is likely that washing is needed, for which you may need a processing licence from your certifying body. On a larger scale, machines exist for topping, tailing and spray washing leeks.

Garlic (All, *Allium sativum*)

Garlic is a high value and surprisingly hardy crop that can be grown as far north as Yorkshire in the UK, provided suitable varieties are chosen. The demand for garlic has been growing and, with good storage, it can be sold for nine months of the year, making it a useful crop for direct sales. A diversity of types is available, ranging from pinky-purple to white-skinned cultivars. Most garlic grown in the UK is soft necked and will not flower. Hard-necked garlic produces a flowerhead that needs to be removed.

Varieties and Seed: garlic is grown from cloves and the seed will arrive as bulbs, which will need to be split into cloves and graded, discarding any diseased, damaged, soft or over-small cloves (less than 1g). Use varieties from northern Europe that are acclimatized to UK conditions. Some varieties can be planted in November for over-wintering, whilst other varieties can be planted in late winter or early spring (ideally by the end of February). Seed can be saved from home-grown stock, providing it is free from virus. Always save the largest top-quality bulbs.

Planting and Growing: a vernalization or cold period of 0–10°C for one to two months is required before planting out. This can be achieved by a late-autumn planting, from October to December or by cold storage. There are also varieties that can be planted in early spring, but no later than early March. Planting can be done manually into marked furrows or through plastic mulch. Machine planting is possible but if cloves are planted upside down then yield and deformities are likely to increase. Rows should be spaced at least 300mm (12in) apart with 150mm (6in) in the row. Plant at least 25mm (1in) deep. The maximum vegetative growth occurs during April and May at around 16°C, until bulbing in June. Some varieties can throw up tall, woody, flowering stalks. These should be cut off when first spotted or they will reduce yield.

Soil Fertility and Rotation: garlic is a heavy feeder that does best in a deep, well-drained friable soil with high organic matter content. The pH range should be between 6–7.5. The tricky period is in early spring, when the soil is not biologically active and nitrogen in the soil has not been mobilized. A derogation may be needed for a fast-acting organic nitrogen-source at this time. Soils that warm easily in the spring are better.

Irrigation: adequate water is needed during the main growth period of late March to mid-June, but irrigation should stop at least two weeks prior to harvest.

Weed Control: as with other alliums, garlic is a poor competitor and, being an over-winter crop, it can often be difficult to carry out mechanical inter-row weed control. Black-plastic or biodegradable mulch is one option, as is mulching with green-waste compost. Post-emergent flaming of garlic may be possible.

Pests: those that affect onions will also affect garlic, but the most serious problems are likely to be stem eelworm and viruses. Clean seed stock is important and control of aphids that may carry the viruses.

Diseases: onion white rot can be devastating and infected land should be avoided. Use of compost may have a suppressive effect. Leek rust can also attack garlic.

Harvesting and Storage: timing of harvest is important, as garlic can double in size during its last month of growth. Garlic can be harvested for fresh use from late June, with the remaining bulbs lifted as the tops start to brown off, usually around mid- to late-July. The outer skin should be tight and the bulbs fully developed. If delayed, quality can deteriorate, with bulbs opening and the skins staining. The bulbs will need to be hand-lifted with a fork or by using an under-cutter, being careful to avoid bruising, which can act as an entry point for disease. High temperatures are needed to cure the garlic for storage. Garlic can be hung to dry or laid out in a polytunnel or glasshouse. After the neck cells have constricted, the leaves can be clipped off using secateurs, leaving 5mm (0.25in) of roots and 10–20mm (0.5–0.75in) of top. When the outer skins are crispy it is ready for storage or sale. Alternatively, they can be made into braids to adorn a farm shop or market stall. Another option, if facilities are available, is to use bulb onion drying techniques, blowing hot air through bulk stores for two to three days. Long-term storage can be in cold stores at 65–70 per cent relative humidity, early types keeping for four to five months and later varieties for six to eight months. The store will need to be checked regularly, at least every month for white mould or other rots.

ROOT VEGETABLES

Root vegetables are the traditional staple of the northern European diet, perhaps because they are productive and bulky crops that store quite well.

They generally come from a range of plant families and care should be taken when planning rotations. The umbelliferas include staple root crops such as carrots and parsnips (as well as celery). They form taproots and consequently do not transplant well. They generally perform well under low-fertility conditions and excessive nutrient supply can make the roots prone to forking or hairiness. Many of the plants are biennials and will set seed if left to grow in a second season.

Potatoes are a major cash crop in temperate vegetable-production systems and represent the family Solanaceae in the root-crop group. In contrast to the other members of this group, they generally do well when placed immediately after the fertility-building phase of a rotation.

Beetroot (Ch, *Beta vulgaris* var. *esculenta*)

Beetroot comes in a variety of shapes, sizes and colours and, with the increased interest in direct sales of vegetables, more unusual varieties are gaining popularity. Different varieties are suited to different uses, such as early bunching or storage. With careful choice of varieties it is possible to produce fresh beetroot all year round, although a good storage variety is essential, as are bolt-resistant varieties for early sowings. The green tops make a good substitute for perpetual spinach.

Types and Varieties: varieties with globe-shaped roots are most common, especially for the supermarket trade and for bunching. Long, cylindrical roots are reputed to have a pleasanter, less earthy flavour however, especially if eaten raw. For the more specialist market there are now white and yellow varieties on the market, as well as one with pink and white concentric rings of colour.

Sowing and Growing: 'seeds' are actually dried fruits containing clusters of seeds rather than individual seeds and so young plants may need to be thinned. Some monogerm varieties are available, although they are not always the most popular choice. The seed needs soil temperatures of 7°C or above to germinate and plants will bolt if subjected to temperatures below 10°C for any length of time.

For early crops, seed can be multi-sown in modules from early February and transplanted at four or five weeks. Most beetroot is direct-drilled in rows at least 300mm (12in) apart. The aim is to have one plant about every 75mm (3in), although thinning is rarely practised. Drilling is usually carried out between April and June, once the soil is reasonably warm. Bolt-resistant varieties can be sown undercover from late January or in the field with protection from March. A reasonably fine tilth is required.

For a continual supply of small tender roots, sow monthly over summer with much closer row-spacing. To encourage the growth of early crops, increase spacing to 150mm (6in) or even 300mm (12in) for transplanted, multisown modules. Crops for bunching can also be raised undercover and cleared before the time to plant summer tunnel crops.

Soil Fertility and Rotation: the beets prefer a rich, loamy soil that is neutral in pH, but are tolerant of most conditions. However, acid soils should be limed. There are no restrictions in the standards for the return of this family to the same area, although it is always good practice to leave a gap of at least three seasons between members of the same family to prevent pest and disease build-up. Possibly the main cause for concern is boron deficiency, which causes unsightly lesions and can result in heart rot. Lesions can eventually cover the whole root, which eventually rots.

Irrigation: beetroot is not a particularly thirsty crop, although prolonged water shortage will cause the roots to become stunted and tough, and sudden fluctuations in watering can cause growth cracks. Overwatering encourages excessive leaf-growth at the expense of root growth and should be avoided.

Weeds: in the early stages one or two passes with a steerage hoe will probably be necessary, although by the time the root starts to swell, the canopy is usually developed enough to shade out most weeds. Passes can be made at reasonable speeds, to throw some soil into the rows and suppress weeds there.

Pests: there are no major pest problems with this crop, although slugs, cutworms and other soil fauna will cause secondary damage if growth cracks or lesions do appear. Rabbits and rodents may gnaw tops off and chew roots, as other food sources become scarcer. Leaf-beet pests, such as aphids, can be a minor problem.

Diseases: common scab (*Streptomyces scabies* and other *Streptomyces* spp.) can be a cosmetic problem, especially in hot dry weather or in light, alkaline soils, although it is not often serious. Crown gall (*Agrobacterium tumefaciens*) may affect beetroot, although again the damage is cosmetic. Violet root rot (*Helicobasidium purpureum*) can affect beetroot, though it is more common on carrots. It is less likely to be seen in the cooler north, although it can be a serious problem in roots stored in clamps or pits. Finally, beetroot black-leg and dry rot (*Pleospora bethe*) is a fungal disease that can be a problem for seedlings, causing the stem base to blacken and shrivel. It is seed borne and often associated with boron deficiency. It is important to promote good seedling development, as they can grow through the most susceptible stages.

Harvest: golf-ball sized roots for bunching can be pulled in July from an April sowing. The maincrop harvest takes place between August and November, depending on variety, weather conditions and sowing date. The crop to be stored is usually pulled in October and November. Plants can usually be pulled by hand and the leaves twisted off or cut at about

30mm (1.25in) above the root. Beetroot will stand a moderate frost and can often be field stored, certainly on well-drained soils, until it starts to warm up in late winter or early spring. It is less tolerant of cold and wet conditions, when rotting tops can spread infection down into the roots.

Storage: bunches with green tops on require rapid cooling, especially in hot weather, and should be harvested early in the day and removed from the field within half an hour during hot summer weather. Even in good conditions they will start to wilt within a day, so are best harvested on the day of delivery. The roots store best at 2°C with about 95 per cent humidity in well-ventilated containers, in clamps or sacks. Field-stored crops may require frost protection, for example, a layer of straw, and their condition will deteriorate, so they are best not left much beyond Christmas.

Carrots (Um, *Daucus carota*)

Carrots are a major crop, widely grown on a field scale and in intensive vegetable-production systems. It can be difficult for small growers to compete with the field-scale producers, but customers of all marketing outlets will undoubtedly want carrots. By choosing varieties for flavour and producing early carrots for bunching, the small grower can create a niche for local direct marketing. Many different types are grown: the short-rooted Nantes types are usually chosen for early production, Chantenay types have short, stumpy, tapered roots with good flavour, Berlicum types are longer and cylindrical, while Autumn King types have large, tapered roots that store well. Hybrids are also available and generally produce better than open-pollinated varieties.

Seed: is available as natural, primed and pelleted. Primed seed has been pre-stimulated to improve germination rate and percentage. This can improve establishment and help gain an advantage over weeds and, so, is particularly useful for the early crops. Pelleted seeds have been covered with an inert substance for ease of precision drilling. At present there is a large number of varieties available as organic seed but few are available as graded seed and, therefore, can't be used in precision drills.

Sowing and Growing: a very early crop for bunching can be obtained by sowing at the end of October/November or January/February in polytunnels. From late January/February the first outdoor crop can be sown under protection of floating polythene or fleece, if soil conditions are favourable. This can help satisfy demand in May and June for freshly bunched carrots, when few UK carrots are available. Sow using a precision drill, if possible in rows 220–300mm (8.5–12in) apart.

Carrots need a fine seedbed, so a minimum of shallow cultivations should be used. Excessive cultivation can cause the soil to slump and cap

Fresh, bunched carrots are popular with consumers

and cause uneven or poor emergence. The system of growing should be geared to the method of weed control that will be used in the crop. Ridges tend to be used on heavier soils and in areas of higher rainfall for easier harvesting. Large carrots will be produced, but with lower overall yield due to lower populations, as rows will be about 750mm (30in) apart. In contrast, the bed system avoids tractor wheelings on the land where the carrots are growing and is suitable for mechanical inter-row weeding with brush weeders or steerage hoes. The row spacing should be tailored to the weeding equipment available. Early carrots should be sown at a lower density than maincrop carrots to encourage faster development of the roots.

Soil Fertility and Rotation: carrots will grow on a wide range of soils, preferring deep, well-drained light loams with a moisture-retentive subsoil and, preferably, free of large amounts of stones. Avoid heavy, clay soils that are difficult to work. The pH range should be 6.5–7.5. Carrots are not a nutrient-demanding crop and can, therefore, be placed later in the rotation than potatoes and leafy vegetables. Ideally a five- or six-year rotation is required to keep in check soil-borne diseases, particularly where the land has been intensively cropped in the past.

Irrigation: in dry conditions irrigation may be required to ensure even germination and/or to ensure a weed flush that can be killed by flame-weeding. Further irrigation can increase yields, but care should be exercised where soils are vulnerable to capping (water little and often).

Weed Control: generally carrots are sown in a single-line drilling or as a multi-line drilling. A higher plant population will be achievable with multi-line drilling, but higher weeding costs will result from the 'dead space' in the bands. Stale seedbeds and pre-emergence flame-weeding are standard cultural operations for organic carrots. As soon as the carrots have emerged and can be seen in the rows, it is common to inter-row cultivate, using a brush weeder, steerage or precision-guided hoe. Some hand weeding will be needed. Research at HDRA has shown that the optimum time for weeding is around four weeks after 50 per cent crop emergence. The use of crop covers can increase weed growth by up to four times and additional weeding may be needed for protected crops. The use of tractor-mounted bed-weeders can reduce the costs of hand-weeding.

Pests: carrot fly (*Psila rosae*) is the major pest of organic carrots, the larvae rendering the crop unsellable by mining into roots and causing dirty tunnels that can be observed near the surface of washed carrots. A combination of crop covers and drilling times can overcome the pest. Carrot flies emerge in May and early June from pupae, laying eggs around young carrot plants, hatching after about seven days and burrowing into the ground to feed on the roots. This is the first generation and normally the highest-risk period. There are three generations of attacks. Crop covers, such as fleece and mesh, can be used during the high-risk period but need to be in place before the flies become active. Populations can be monitored using sticky traps and forecasting services are available from the HDC. Maincrop carrots sown in late May or early June can miss the first generation but, in some years and areas, can be affected by later generations. Early harvest can avoid damage from late generations. Other pests that could be encountered include cutworms (caterpillars of several moth species) and wireworms (the larvae of the click beetle), particularly after grass. Cutworms are active from mid June (prediction services are available) and can be managed by irrigation. Monitoring of wireworm activity is necessary if there is a risk. Soil analysis is available, baiting kits can be used or potatoes can be buried in the soil a few weeks ahead of intended drilling and examined for damage.

Aphids (*Cavariella aegopodii*) are occasional pests that might cause damage in some years, but are better known for transmitting motley dwarf virus, which causes a reddish tinge to foliage and can reduce yield.

Diseases: *Alternaria* blight is a seed-borne disease and ensuring the use of healthy seed is vital. *Alternaria dauci* causes yellowing of the leaves and can reduce quality of tops for bunching and yields for maincrop carrots. The presence of *Alternaria radicina* can cause severe losses at the seedling stage. Cavity spot (associated with *Pythium*) is the most serious disease of conventional carrot production but has been less of a problem in organic crops, probably due to longer organic rotations. Fields should be selected that have no history of cavity spot problems – soil tests are available if this is not known. Varieties are available with good resistance. Early lifting can also limit symptom development. Violet root-rot (*Helicobasidium purpureum*) is a soil-borne fungus, affecting many root crops, that can be contained by crop rotation. Carrots are the most susceptible crop and develop purplish fungal growth on the root surface. When spotted, prompt harvest is recommended to prevent build-up. It can persist through volunteers and perennial weeds, such as bindweed, so good weed control is necessary. Viruses such as carrot motley dwarf virus and parsnip yellow fleck virus can be spread by the carrot willow aphid.

Harvesting: is much easier on the lighter soils. Carrots should be harvested as soon as they reach the required size for market. Care is needed, as they can be very susceptible to mechanical damage, some varieties more than others. Bunching carrots can be undercut or hand-lifted and transported inside for washing, bunching and cooling. For maincrop carrots, harvest in October and November when the crop is mature and leaves are yellowing; the crop can be harvested by undercutting for hand-harvesting, top lifters (lifts the crop by foliage) or share lifters that lift the carrots onto an elevator and separate the soil. The carrots will need to be topped first if using a share lifter. It depends on the market requirement as to whether washing is needed. The tendency is to move to washed carrots, though they will not keep as long. Crops for storage should be harvested in good weather and soil conditions, to reduce the amount of soil and plant debris brought into store.

Storage: carrots can be lifted and cold stored for a limited period and will store best if not washed, cooled rapidly and kept in high-humidity conditions. Carrots can be stored in the ground, provided harvesting will not damage soil structure. They should be protected from frost and may need protecting from carrot fly damage by fleece. In this way they can usually be kept until about Christmas. Longer-term storage in the field is possible on lighter soils. They can be ridged up with soil or covered with black plastic and a thick layer of straw. This can be a problem for removal or incorporation, as it can cause nitrogen lock-up. Clamping or cold storage is another possibility, but crops need to be inspected to ensure no damaged or infected roots go into store. They can be cold-stored to last beyond March (until about June), when field-stored carrots will deteriorate, but there will be a loss of skin quality. This will not be as much of a problem for the 'dirty' crop.

Parsnips (Um, *Pastinaca sativa*)

The parsnip is a native of Britain and more popular in the UK than in many other countries. It is a good winter vegetable that can be stored in the field and harvested over a long period.

Seed and Sowing: pelleted seed can be used for precision drilling. Fresh seed needs to be used each year. They are usually grown on a bed system with a row spacing of 450mm (18in), precision drilling to 50–100mm (2–4in) apart. Market specifications will determine the spacing. Although some books may suggest drilling from January onwards, they are much better drilled when soil conditions are warmer and establishment is quicker in April or early May, unless a particularly early crop is required.

Soil Fertility and Rotation: while parsnips are able to grow on heavy, clay soils they prefer light, well-drained loams that have not been freshly manured. They are, therefore, better at the back-end of the rotation. The pH should be between 6.5–8.

Irrigation: may be necessary for establishment and later for yield improvement.

Weed Control: stale seedbeds and pre-emergent flame-weeding should be used. The later sowing will enable more of a weed strike. On a small scale, radish can be drilled within the parsnip rows to mark them for weeding and removed as a crop while the parsnips are establishing.

Pests: as with carrots, the main pest is likely to be carrot fly (*Psila rosae*), though damage is not usually as bad and parsnips are seldom grown under fleece or crop covers.

Diseases: although a number of foliar diseases can affect parsnips they are rarely serious and the biggest problem is likely to be canker (*Itersonilia pastinaceae*). Large lesions on the crowns and sides of the roots are caused by several different pathogens and can get worse as the season progresses. Some of these can be seed-borne and only healthy seed should be used. Other canker pathogens are soil-borne and can be contained by good crop rotation. Some varieties, such as Javelin and Arrow, have good resistance to black canker. Methods of cultural control include keeping the crowns well covered by soil and minimizing root damage during weeding and harvesting. Large roots are more affected than smaller ones. Regular monitoring is advised, so that crop harvesting can be accelerated if problems occur. Powdery mildew (*Erysiphe heraclei*) can be found from July onwards, but does not normally cause significant damage.

Harvesting and Storage: can commence when roots meet market requirements, though flavour will improve if they can be left until touched by frost in October. The crop is hardy and the crop can be left in the ground and lifted as required until the end of April, when they can start to re-grow and become woody. Top pullers may be used, before the leaves senesce, or a share lifter. On a smaller scale an under-cutting bar or hand-lifting with a fork are options. They can be lifted and cold-stored if the ground is needed. Canker is a problem or damage to the soil will occur if left to be harvested in poor winter conditions. They will store for two to three weeks in a clamp and up to six months in cold storage at 95–98 per cent humidity. Once parsnips are washed they deteriorate rapidly and need to be harvested, washed and packed on the same day, to reach the shop without discolouring. If it is possible, it is best to sell them dirty.

Potatoes (Sol, *Solanum tuberosum*)

Potatoes are grown organically for a wide range of markets. Where soil conditions are suitable, they are a popular crop to include in an organic rotation, as they can be relatively trouble-free to grow. There are a number of marketing opportunities, ranging from new and early potatoes through to maincrop types, as well as salad and speciality types (such as heritage varieties).

Seed Quality: organic potato seed is readily obtainable and there is generally a much wider range of varieties available than with many other crops. As with all crops, good-quality seed is an essential starting point. This is particularly important for potato seed tubers, as they are prone to transmitting diseases. In choosing potato seed there are a number of factors to consider.

Seed size: size of seed determines the number of stems produced by a single seed tuber. A single, large seed tuber will produce more stems than a small seed tuber. However, when purchasing seed on a weight basis, a given weight of small seed will produce more stems than the same weight of large seed and can therefore be used to plant a larger area. This is reflected in a price premium for small seed. Seed that has been graded into a narrower range of sizes (say 30–35mm (1–1.5in)) will have less variability in number of stems per seed tuber, so will produce a more even distribution of stems in the field. This is more important for early crops, especially if the grower is aiming to produce tubers for a specific size grade, such as for punnets. The practice of cutting seed to make it go further is definitely not recommended, as it greatly increases the chances of bacterial infections, such as blackleg (*Erwinia carotovora*).

Sprouting: most seed will break dormancy after February and begin to produce sprouts, unless stored at a cold temperature. Organic growers often use sprouted seed to advance the emergence, growth and harvest of the crop, reducing the chance of late blight infection. Seed should be sprouted in trays in the light and kept frost-free and well ventilated. If sprouted in the dark, brittle sprouts are produced that are likely to break off, leading to delayed and uneven emergence when planted.

Seed diseases: selecting disease-free potato seed is paramount, as using infected seed not only risks infecting the current crop but can introduce diseases to the soil that can persist for years. Seed used should be of at least the certified (CC) grade if growing for ware use and at least basic grade (AA, Elite or Super Elite) if growing for seed. More information about requirements of grades is available from DEFRA. Home saving of seed is not generally recommended in organic systems, as this greatly increases the chances of spreading infection. There are a number of diseases to look out for when selecting seed. These include black scurf (*Rhizoctonia solani*) and skin spot (*Polyscytalum pustulans*), two diseases which potentially have the most serious consequences, as they not only infect progeny tubers, but also affect the growth of the crop. Black scurf is often transmitted to the stems of the crop and is visible as brown lesions (stem canker). In severe cases the lesions can kill off individual stems, resulting in proliferation of large numbers of secondary stems. The consequence of this is delayed crop development and wider variability in progeny tuber size. Skin spot can cause death of the eyes of the seed, preventing crop emergence in extreme cases.

Varieties: a wide range of varieties has been tested under organic conditions by NIAB and HDRA. As with conventional potatoes, it is essential to select the appropriate variety according to the time of harvest and the market you are aiming for. However, selecting varieties for resistance to blight is a much higher priority when growing organically, as control measures are more limited. Early varieties that may be suitable for organic production include Premiere, Swift, Maris Bard, Junior and Orla. Of these, Orla and Premiere show the best resistance to blight. Cosmos and Milva have been grown successfully as second earlies, with only low levels of foliage blight and very little tuber blight. A wide range of varieties are available for growing as maincrop, including Appel, Cara, Lady Balfour, Romano, Sante and Valor. Cara shows particularly good resistance to tuber blight, but is one of the latest-maturing varieties and produces a very large canopy, making defoliation difficult. New Sárpo varieties are now becoming available that show very good resistance to foliage blight, but not all show good resistance to tuber blight. When growing varieties on heavier soils, slug resistance may be an important factor determining varietal choice. Of the above varieties, Romano shows the best resistance, with Sante and Lady Balfour also showing some limited resistance.

Planting and Growing: the ground should be ploughed and power harrowed to produce a seedbed to a depth of 200mm (8in). The soil should be clod free, especially if the crop is going to be harvested using a potato harvester rather than a spinner. In this case it may be necessary to use a clod separator and de-stoner, to avoid too many clods and large stones being harvested. Organic potatoes are more commonly grown in ridges rather than beds, in order to facilitate weed control by re-ridging (that is, knocking down and building up ridges). Ridges should be firm so that tubers are covered, to prevent greening and to reduce the chance of blight spores infecting the tubers. Spacing between the ridges can vary from 700–900mm (28–36in) to fit in with the row-width of the farm. It is difficult to make general recommendations for spacing within the rows, as it is dependent on many factors, such as seed size, variety and the size market for which the crop is being grown. Target spacing tends to be wider for organic crops, to facilitate air movement within the canopy, reducing the chances of late blight. As an approximate guide, spacings will be in the range 200–300mm (8–12in) for early season crops and 250–400mm (10–16in) for maincrops. Larger seed should be planted at a wider spacing, as should varieties that produce a large number of tubers. Where growers are aiming to produce smaller tubers for the punnet market (35–50mm (1.25–2in)), crops should be planted at a closer spacing, whereas, if a higher proportion of bakers are required, then a wider spacing should be used.

Soil Fertility and Rotation: certifying bodies will not permit growing potatoes for more than one year in four in a rotation. Ideally, the interval should be longer than this, especially if there is a problem with nematodes in the soil. Potatoes are a nutrient-demanding crop and are, therefore, frequently grown directly after the fertility-building phase. Up to 25t/ha (11 tons per acre) of animal manure, which should be composted or well-rotted, can be applied before planting. Although many growers like to ensure that enough nitrogen is applied to produce a dense lush-green canopy, this is not always desirable in a potato crop. In early crops, applying too much nitrogen can delay maturity and excessive foliage growth will make defoliation difficult. A large, dense canopy will also create a microclimate favouring blight infection. Extremely indeterminate varieties, such as Cara, that produce large amounts of haulm-growth benefit from being grown at lower levels of nitrogen than other varieties. They could possibly be grown in the second phase of the rotation, especially on highly fertile fen-peat soils. It is important to ensure that levels of potassium, phosphorus and minor nutrients, such as manganese and magnesium, are adequate.

Irrigation: although potatoes are grown without irrigation on some sites, on lighter soils and drier areas, it should be considered essential in order to achieve decent yields and quality. Tradition is to put on 'an inch a week', but there are a number of ways that the amount of water applied to crops can be targeted more effectively. This is particularly important as restrictions on water use become more rigorous. Judicious application of water is also important in organic systems to reduce leaching of nutrients. A number of organizations use computer models to schedule irrigation. Most of these calculate crop water use using local meteorological data and crop cover, then, in combination with information on soil type, estimate the amount of water held in the soil that is available to the crop. This service is reasonably accurate for scheduling water use for yield, but is not so good at predicting precise moisture levels in the ridges for scab control. Other services measure the amount of water in the soil, using equipment such as neutron probes or enviroscan. These have the advantage of actually taking a direct measurement of soil moisture content, but it must be borne in mind that the measurements are taken from a single point within the field that only represents a very small volume of soil. Adequate replication is vital to ensure that the measurements give a realistic picture of overall soil moisture levels in the field. As mentioned in the section on diseases, irrigation is one of the main means of preventing common scab. Control is achieved by restricting the soil moisture deficit during the period between tuber initiation and up to four weeks afterwards.

A number of different types of irrigation are used on potatoes. Mobile rain-guns that are drawn across the field by a hose reel are the most popular for large areas, but water application is uneven and wasteful. Sprinkler systems are often used on smaller areas. The disadvantages of overhead irrigation systems are that water is lost through evaporation from the canopy and run-off from the soil surface. Applying water to the canopy also increases the risk of blight infection. Drip systems are by far the most efficient method of applying water and reduce the risk of blight, but this must be weighed up against the cost of the tubing and the labour involved in setting up the system. Labour costs will be greater when reusing the system, as drip lines will not be the correct length for a different-sized crop and, if drip tape is reused, it will almost certainly require repair.

Weed Control: weeds are not normally a serious problem and potatoes are often considered to be a 'cleaning crop' for weeds within the rotation. The most troublesome weeds are likely to be perennial grasses, such as couch. Fat hen produces tall plants that can penetrate a potato canopy and can be problematic. Most weeding operations will be carried out in the early stage of the crop. During the later stages of the crop, the canopy should be vigorous enough to compete against most weeds. Prior to emergence, thermal weed control may be used. The ridging body is the most common tool used for control of weeds in potato crops after emergence, although chain harrows and rolling cultivators may also be used for inter-row cultivations. These operations are best carried out when weed seedlings are small and unlikely to re-establish. The number of passes necessary depends on the weed density and the quantity of potato foliage.

Pests: a range of pests will attack potato crops, many of which can be controlled by cultural methods or by selecting a more resistant variety where there is a known problem. Pests include:

Wireworm: these thin, orange 'worms' are the larva of the click beetle. They bore small holes in the tubers and can render large proportions of the crop unmarketable. They tend to build up after a long grass/clover ley, which, unfortunately, is when potato crops are most commonly grown in the rotation.

Cutworm: another pest that can do extensive damage to potato crops. Cutworms are smooth, pale-brown caterpillars of the turnip moth (*Agrotis segetum*). If the ground is very dry at emergence, cutworms can kill whole plants by chewing the stems as they emerge. Later in the season they can render a large proportion of the crop unmarketable by boring large holes and hollowing-out tubers. Damage is most prolific under hot, dry conditions and the severity of attack can be greatly reduced by irrigating.

Slugs: can be a major problem when growing potatoes on heavier soils in wet conditions. Numbers tend to build up after long grass/clover leys. Although biological control using nematodes exists, this option is presently too costly to implement on a field scale. As numbers tend to increase after August, earlier harvesting can reduce the problem, perhaps choosing a second early rather than a late maincrop variety in such cases. Of the red varieties, Romano shows the best resistance to slugs. Unfortunately, at present, there are no white maincrop varieties grown organically that are as effective against slugs.

Aphids: do not normally cause extensive damage to crops directly unless infection is severe, but they can transmit a number of viruses. The most common aphid feeding on potato crops is the peach potato aphid (*Myzus persicae*). They are of most consequence in seed crops, where maintaining virus-free material is paramount. The most effective measure to minimize the risk of transmission is to grow the seed crop in an area well away from any ware crops. Aphid populations in organic crops can be reduced by taking measures to encourage predatory insects: planting beetle banks or attractant flowers, such as *Phacelia*, will help boost predator numbers. Large areas of crop should be broken up by strips to improve predators' dispersal. In extreme cases, the crop can be sprayed with potassium soap, although this may also harm predators.

Nematodes: eelworms are microscopic animals that live in the soil. A wide range of species attack potato plants, but the most economically damaging is the potato cyst nematode (PCN), which can cause large yield-losses. *Globodera pallida* and *Globodera rostochiensis* are the most significant species attacking potato plants. They form small yellow (*G. rostochiensis*) or white (*G. pallida*) cysts on the roots that are only just visible to the naked eye. Symptoms in the field are stunted plants with sick-looking foliage that wilts easily. These symptoms are commonly distributed in patches around the crop. The only effective way of preventing a build up of nematodes is to ensure that there is a long enough break between potato crops – six years minimum, or longer if nematodes are known to be a problem. Soil samples can be sent to a number of laboratories for analysis to identify the species and population levels within the soil. It is important to follow the laboratory's instructions for taking samples. Samples need to be taken from a large number of places within the field, as nematodes are distributed within discrete patches that can easily be missed. Choosing resistant varieties will also reduce symptoms. Some varieties, including Cara, Lady Balfour, Sante and Valor show good resistance to *G. rostochiensis*, with Sante also showing partial resistance to *G. pallida*. Lastly, good sanitation practices, such as ensuring that cultivation equipment, footwear and vehicles are clean, will reduce the risk of nematodes being spread between fields.

Diseases: foliar diseases are the most apparent diseases in the growing crop.

*Late blight (*Phytothera infestans*)*: this is by far the most serious disease of potato crops and has the potential to destroy entire crops very quickly. The first visible signs appear as dark spots on the foliage with a characteristic, pale surround. In damp conditions this can rapidly spread throughout the crop. Rainfall will wash spores from the leaves onto the tubers, which will also rot, either in the soil or when put into storage. There are a number of measures that can be taken to minimize the impact of this disease in organic potato crops. The first is to select a variety with some resistance to the disease. These are listed in the variety section, but lists are also available in the NIAB variety listings (*see* Useful Addresses). Unfortunately, some varieties that are very popular with the major retailers, such as Nicola, also happen to be very susceptible to the disease. Early planting and advancing the crop by chitting the seed will also reduce the risk, as chances of infection increase later in the season. The British Potato Council (BPC) provides a warning system for growers that can predict when there is a high risk of blight occurring. Taking local climatic data, it uses the system of 'Smith Periods', a period of two days when temperatures are not less than 10°C and relative humidity is above 90 per cent for at least 11 hours each day, to predict when risk of infection is high. The system can alert growers through email or SMS text message. It is important for the grower to inspect the crop regularly for signs of blight. If it is present in a few isolated plants, these should be destroyed to prevent further infection. Risk of further infection is greatly reduced if subsequent weather conditions are warm and dry. If infection is widespread, it may be necessary to defoliate the entire crop. This is particularly important if rain is forecast, which greatly increases risk of infecting the tubers. The grower must decide on the balance between losing yield by defoliating early or risking losing the entire crop through blight infection. Defoliation can be achieved using a propane-powered burner and may take more than one pass, depending on the condition of the canopy. A foliage topper can also be used, but will not kill blight spores. It is important to ensure that the entire canopy has been dead for at least two weeks before harvesting, to minimize risk of infecting the tubers. It also allows any rotten tubers to be identified and separated from the marketable fraction of the crop.

Growers may also currently apply for a derogation to spray with copper products, which have some effect in delaying the onset of blight. This should only be used when risk is high and not as a prophylactic measure. At present a maximum of 6kg of copper per ha per year (5.5lb per acre per year) is allowed and its use is likely to be banned in the future, amid concerns over its adverse effect on soil biological properties. Good sanitation around the farm is very important in minimizing spread of the disease. Dumps of potatoes that are allowed to sprout are a key source of infection and should be destroyed or at least covered with polythene. Volunteer potatoes in other crops are also a source of infection and should be weeded out.

Blackleg (Erwinia carotovora): this bacterial disease is transmitted primarily on the seed and can affect both the growing crop and tubers in storage. Early symptoms in the field are wilting of the leaves, followed by blackening, then collapse of the lower stem. The disease can be transmitted via stolons to the tubers, causing a pungent-smelling soft-rot. Infection in store can be reduced by ensuring conditions are dry and well-ventilated. Although bacterial levels in seed tubers can be tested, this will not necessarily give a realistic indicator of infection at the time of planting, as levels can fluctuate considerably during storage.

Skin-blemishing diseases are potentially very damaging in the crop, as marketing requirements become more stringent. Skin diseases include:

Black scurf (Rhizoctonia solani): this is a fungal disease visible as raised black-flecks on the tuber surface and is of little consequence at low levels in ware crops. However, it is more of a problem in seed crops, as the disease can cause lesions on the stems of the subsequent crop. In severe cases the lesions can kill off individual stems, resulting in proliferation of large numbers of secondary stems. The consequence of this is delayed crop development and wider variability in progeny tuber size. The disease will also cause blemishes to the progeny tubers.
Skin spot (Polyscytalum pustulans): this fungal disease has important consequences for seed tubers, as it can infect the eyes and prevent them from sprouting. The symptoms are visible as pimples, which are often surrounded by a dark sunken ring.
Common scab (Streptomyces scabies): a skin-blemishing disease caused by actinomycetes, soil bacteria that form filamentous chains. It appears as corky, angular, brown scabs on the tuber surface that can render crops unmarketable. It is most prevalent on freely draining soils and the most effective method of control is ensuring the soil is moist until four weeks after tuber initiation. This is often incorporated into irrigation scheduling-programmes that aim to restrict the soil moisture-deficit during this period. The disease is also favoured by high soil pH, so it is best to avoid applying limestone before planting potatoes. As the disease can be transmitted on the seed, using clean seed will also help reduce infection, although it can also remain in the soil for many years.
Powdery scab (Spongospora subterranean): forms scabs on the surface of the tuber that can erupt and produce a mass of spores. It can sometimes be confused with common scab. In extreme cases it can lead to deformed-shape tubers. It is most prevalent on cold, damp, heavy soils but also occurs on lighter soils when irrigated. As it can be transmitted through infected seed, it is important that clean seed is used. It can also be

transmitted in animal manure, so infected tubers should not be fed to livestock if the manure is going to be spread on potato-growing land. As with common scab, powdery scab is also transmitted through the soil and can remain there for many years.

Potato also suffers from a range of viral diseases. Of the viruses that infect potatoes, potato virus Y (PVY or leaf drop streak) is the most common. Infection is most commonly introduced into the crop by the peach potato aphid (*Myzus persicae*). Primary symptoms then develop that are initially visible as small, dark flecks on the leaves and streaks on the veins. As the condition progresses, leaves shrivel and drop off. Symptoms frequently occur on isolated stems where the aphids have been feeding. Where infected seed is used, secondary symptoms occur, producing severely stunted plants with wrinkled leaves. The first step to preventing this disease is the use of healthy seed. Basic seed should have negligible levels of the disease and certified seed will have low but acceptable levels. It is most important to keep the disease out of seed crops. Maintaining distance from other potato crops will reduce the risk of infection, as this virus does not persist in the aphid for long. Crops grown for seed should be tested for virus. Individual leaves can be tested in the field using a laminar-flow antibody kit to give an indication as to whether the disease is present. Tuber samples should also be sent to a laboratory to test for the presence of virus. A number of organizations will offer this as a commercial service to seed producers. Selecting resistant varieties is another line of attack, as many of the varieties in use today show some resistance to PVY, including Cara, Lady Balfour, Romano and Sante.

Other viruses include leaf roll virus, which is the second most common potato viral disease and is also carried by the peach potato aphid. Although it takes several hours of feeding for the aphid to acquire the virus, it persists in the aphid for a long time and is easily transmitted to other plants. Symptoms include rolling of the leaves on upper parts of the plant. Secondary symptoms of plants grown from seed appear similar. Again, the best precautions against this disease are isolation, use of clean seed and choice of variety. Of the varieties commonly grown organically, Sante shows good resistance, whereas Cosmos, Lady Balfour and Romano are more susceptible.

Spraing is a disorder of potatoes that appears as brown crescent markings on the flesh of the tuber and can render crops unmarketable if incidence is high. It is most commonly caused by the tobacco rattle virus (TRV), which is transmitted by free-living nematodes. There are often no foliar symptoms and the condition can only be identified by cutting open tubers. Spraing may also be caused by potato mop top virus (PMTV), which is transmitted by the powdery scab fungus. Other symptoms of PMTV may be visible on the plant, including stunting and bright-yellow blotches on the leaves. Use of clean seed, rotation and choosing a resistant variety are all ways of preventing this disorder. Many varieties show moderate resistance, with Romano showing slightly higher resistance and Cara being moderately susceptible.

Defoliation and Harvesting: early potatoes will need defoliating before harvest. This can be done either with a burner or a topper. Although the topper is cheaper and quicker, it is more likely to transmit any blight spores to the tubers if the disease is present. If loose-skinned tubers are required, harvest after defoliation. Set-skinned tubers are generally more popular, especially amongst the supermarkets. The skin-setting process is initiated by defoliation, after which tubers will need to be left in the ground for at least two weeks for the process to complete. A small sample should be dug before harvesting to check for skin set. Harvesting of maincrop ware-potatoes should follow the same procedure as early potatoes. If the canopy has senesced, then some degree of skin set will have already occurred, so tubers will not need to be left in the ground for so long. Good skin set is essential if the crop is to be stored, as it reduces the chances of infection entering the tubers. Processing crops should be harvested when mature, so that dry matters are high enough to fry properly. The most popular varieties, such as Hermes and Lady Rosetta, which are grown conventionally for processing, tend to be very susceptible to blight and so are difficult to grow organically. Alternatives such as Agria are popular for organic processing.

When growing crops for seed, the appropriate guidelines for the seed class should be followed for burning off the crop following the final

Defoliating potatoes before harvest

inspection. Regulations for harvesting and details of disease and rogue tolerances in seed crops are laid out in the seed-potato classification scheme leaflet available from DEFRA.

Tubers can be harvested by hand, by using a spinner or using a potato harvester, depending on the scale of the operation. Harvesting on a rainy day should be avoided and the temperature should ideally be between 8–25°C. Care should be taken to avoid bruising. Avoid drops of more than 200mm (8in) onto the metal surface of a trailer and more than 600mm (24in) onto other tubers. Susceptibility to bruising is increased markedly under cold conditions and sometimes after very dry seasons. Agria is quite susceptible to bruising, whilst Cara and Cosmos are fairly resilient. It is important that the harvester is set correctly, so that all tubers are removed from the field. Tubers left in the field will pose a volunteer problem in subsequent crops and are a source of blight infection.

Storage: the first important point for successful storage is to ensure that the material entering store is of good quality. This entails removing damaged tubers, rots and greens. Firm skin set is also essential. It is recommended that potatoes undergo a curing period of two weeks at 15°C prior to being reduced gradually (ideally 2°C per week) to the storage temperature. The simplest method of storing potatoes is in a barn without refrigeration. Potatoes can be stored for a number of months, providing that temperature is not too high but they are also protected from frost damage. This can be achieved by using a wall of straw bales around the clamp. Dry, well-ventilated conditions should be maintained using ducts. Light should be excluded to prevent greening.

As organic potatoes must be stored without the use of chemical sprout-suppressants, some form of refrigerated storage will be necessary for producers wishing to store potatoes for longer periods into late winter and early spring. A cold store requires considerable investment, so smaller growers may opt to use a storage company to store their potatoes. The company would need to be certified by an organic certifying body. The store should be sealed, but provided with ventilation and the capacity for drying. After the two-week curing period at 12–15°C, the temperature should be reduced to 3°C for ware potatoes and 8°C for crisping potatoes. Crisping potatoes that are stored at a lower temperature will produce an unattractive dark colour when fried.

Storage Disorders: a range of disorders can develop during storage and some of the most common are highlighted here.

*Dry Rot (*Fusarium solani*)*: this rot appears with a wrinkled skin around the infected area, which contains pustules. The fungus can be transmitted by soil or seed and commonly infects tubers that are mechanically damaged during harvesting. As the disease is favoured by warm, humid, storage conditions, storing in a cool, dry place will reduce incidence of this disease.

Gangrene (Phoma fovea): sunken, discoloured areas appear on the outside of the tuber and internal tissue is blackened with a distinct margin. As with dry rot, the fungus can be soil- or seed-borne. Symptoms often do not appear until being stored into December or January. As the disease is encouraged by cool temperatures, a two-week period of curing at 12°C at the beginning of storage will promote wound healing, reducing the chances of infection.

Silver Scurf (Helminthosporium solani): this is a tuber-blemishing fungal disease that develops in storage. Symptoms appear as dull, silver patches on the tuber surface. It is primarily transmitted by infected seed and conditions are favoured by warm, humid conditions in storage. The disease can be prevented by using clean seed, dry curing and storing at low temperature.

Other tuber disorders, such as sprouting and greening, can be prevented by controlling storage conditions. Sprouting of tubers will take place if the temperature is too high. It is more likely to be a problem in processing crops stored at 8°C than ware crops stored at 3°C, but there is little that can be done to avoid this, as lowering the temperature will have adverse effects on processing quality. Greening can be avoided by excluding light from the store.

Turnips (Br, *Brassica campestris* var. *rapa*) and Swede (Br, *Brassica napus*)

Turnips and swedes are quick-growing brassicas, often used as a catch crop. The main distinction made between swedes and turnips is that the former are a winter crop, whilst the latter are a summer or autumn crop, although they are distinct genera, with swedes probably representing a cross between turnip and kale. They can be grown under protection for an early, bunched crop. They are a relatively minor crop, but can be useful for adding diversity to a box or stall in spring. They can also be grown in succession through to the autumn.

Sowing and Growing: turnips can be drilled in January or February under protection with the first outdoor sowings in March, sowing successionally every fortnight until May, if required. Maincrop turnips can be sown between mid-July and late August. Swedes are direct drilled, with early crops sown in April and main crops sown in May to June. Spacing between rows is 450mm (18in). Spacing within rows can range from

100mm (4in) for early harvested turnips to 250mm (10in) for maincrop swedes. On a small scale, swedes are sometimes raised in modules and transplanted, to try and avoid the period of maximum cabbage root fly activity, but root shape and yield can be impaired.

Soil Fertility and Rotation: Unlike other brassicas, swedes do not require a highly fertile soil and should not be grown in a soil that has had muck recently applied. Boron deficiency can manifest itself in swedes as an internal browning of the root.

Irrigation: should not normally be needed.

Weeds: as they are usually a directly seeded crop, early weed control is more of an issue. Drilling should take place into a stale seedbed that is flame-weeded shortly before emergence, then subsequent weed control achieved using brush weeders and steerage hoes.

Pests: cabbage root fly is the most economically damaging pest in turnips and swedes, as feeding activity scars the surface of the root, spoiling its appearance. In swedes, a small amount of cabbage root fly activity can be tolerated, as most scarring occurs at the bottom of the root, which can then be trimmed off. However, it can seriously reduce the saleability of small turnips that are to be sold intact, and such crops are commonly grown under fleece as protection. For the bunched turnip crop, leaf quality is important and this can be seriously damaged by flea beetles eating small holes into the leaves. This should not be a problem in a tunnel, but outside sowings may need to be covered by fleece or mesh. Slugs can also be a nuisance for the early crop.

Diseases: are not generally a problem, although some of the more common brassica diseases may be seen.

Harvest and Storage: early turnips can be harvested eight weeks after sowing and sold in bunches with the leaves still on, if the foliage looks respectable. Later crops take three months to mature and roots can be lifted from late September onwards. They are less hardy than swedes and may need protecting from frost by fleece or straw if being harvested from the ground. Early swede crops can be harvested from July to September for early varieties and September to March for main crops. Trimming of the tops and root bottoms will be necessary, removing any cabbage root fly damage if it is present.

PERENNIAL VEGETABLES

Perennial vegetables can remain in the ground from one year to the next. This should be borne in mind, as any area put down to these vegetables will be unavailable for other cropping for some period of time. Weed control over the long cropping period is likely to be the main challenge once these crops are established.

Asparagus (O, *Asparagus officinalis*)

Asparagus is a perennial crop that can remain in the ground for more than ten to fifteen years. There is a good demand for fresh local asparagus in season, but it is already a premium crop and it remains to be seen how much of an organic premium on top of this consumers will pay. It is considered a delicacy, as it is only available for eight weeks of the year. The main challenge for growing organic asparagus is weed control.

Planting and Growing: asparagus is more readily grown from crowns than raised from seed. Common varieties are Gijnlim (an early variety producing from mid-March to June) and Theilim (a later variety). Most modern varieties are supplied as all-male one-year-old crowns. On receiving crowns they should be planted as soon as possible, as they deteriorate fairly quickly. For planting out in spring, dig trenches 200mm (8in) deep with 600mm (24in) between the rows. Form a 50mm (2in) mound within the trench and place in the spears at 300mm (12in) spacing. Replace the soil in the trench. Leave the spears and ferns to grow in the first season without cutting them to allow the crop to establish well. The foliage will die off in the autumn and this should be cut back. In the second season a light harvest can be taken from the spears, but it is best to allow continued growth so that the crop can establish properly for harvest in the third.

Soil Fertility: asparagus is best grown on sandier soils. Acidic soils will require limestone before planting. Manure applications can be made in autumn, after the ferns are cut down.

Irrigation: it is important that the soil is kept well irrigated, especially on free-draining soils.

Weed Control: this is the main challenge of growing a perennial organic crop. A number of approaches can be employed, including mechanical weeding, hand weeding or use of mulches. Mechanical weeding is not possible within the row during the period of spear production in spring. It is essential that the site is free of perennial weeds before planting, as these will be very difficult to remove later in the season. Shortly after planting, when plants are established, some mechanical weeding within the rows can be carried out by dragging flexible, spring-mounted tines through the crop. Although this may cause some damage to the ferns, it will give the ferns a chance to outcompete the weeds. Once ferns are larger this will no longer be possible, as it will cause too much damage to the crop. In autumn, the fern is cut down and the soil is lightly cultivated, then ridges drawn up. This allows for some weed control between seasons.

Use of dying mulches is one technique for weed control that has been applied with varying degrees of success. Winter rye is sown between the rows in the spring. The spring sowing deprives the plants of the normal, cold, vernalization stimulus needed for flowering and the rye plants die down naturally, forming a mulch. The effectiveness of this is dependent on achieving the right balance between sowing the rye early enough to out-compete the weeds, but not so early that there is the risk it could become vernalized and go to seed. Living, low-growing mulches, such as clover, have been investigated but in many cases have caused severe losses in yield through competition. Other approaches that have been tried include flame weeding, hand weeding and the use of geese.

Pests: asparagus beetle (*Crioceris asparagi*) is the main pest and likely to make an appearance in crops. It is easily recognized by its characteristic black-and-yellow markings. During the harvesting period it can feed on the spears, reducing quality, and also defoliate ferns, reducing yield. It can be controlled by eliminating fern residues after autumn to remove over-wintering sites. Ladybird larvae, predatory chalcid wasps and chickens can act as effective predators. Other pests include asparagus aphid (*Brachycolus asparagi*), which feeds on the spears causing stunted, distorted growth known as 'witches broom'. As with other aphids, control can be achieved by encouraging predators or, in extreme cases, spraying with potassium soap.

Diseases: fusarium wilts (*Fusarium oxysporum, F. moniliforme*) are seen as yellowing of the foliage, rotting of the spears and wilting of the foliage. Severity of the disease depends on the varietal susceptibility and the season. In most well-managed crops, fusarium wilt should not be a major problem. If the disease becomes severe, it may be necessary to remove the entire crop and not grow asparagus on the same site. Asparagus rust (*Puccinia asparagi*) symptoms normally appear after the cutting season on the needles as orange spots and, later, as brick-red pustules. In severe cases it can result in subsequent yield-losses by prematurely defoliating the ferns before the autumn. It can be controlled by growing less susceptible varieties and by planting spears at a wider spacing, in order to reduce the humidity within the canopy. Needle blight (*Cercospora asparagi*), like rust, can cause yield losses in the subsequent season by premature defoliation of the ferns. It occurs from June onwards under conditions of high humidity and symptoms appear as grey spots with a thin purple margin. It can be effectively controlled by removing the crop residue.

Harvesting: the first two seasons should be devoted to establishing the crop, so the first major harvest is in year three. Spears should be cut just below the soil surface, starting in March for early varieties and finishing in June for late varieties. Spears should be cut before they open up and become unmarketable. It is an easily perishable crop, so should be stored at low temperatures and marketed as soon as possible. At temperatures of 2°C it should store for ten days, if kept well-ventilated and at high humidity to prevent drying out.

Globe Artichoke (O, *Cynara scolymus*)

The globe artichoke is not to be confused with the Jerusalem artichoke (*see* below) and is a perennial thistle grown for its edible flower buds. The artichoke flower consists of a cone of short, thick-stemmed bracts (or leaves) that are tender and edible. They are attractive plants and can also be a good food source for birds if the flower heads are left on over winter. They can also be sold as fresh or dried flowers. It has limited demand, but it could be a useful crop for the restaurant or farmers' market. They are not normally cropped for longer than three years and, in order to obtain a steady supply of heads, fresh beds should be planted each year.

Seed/Transplants: they can be grown from seed or from slips or suckers. Plants from seed can be variable and not always true to type, but hybrid varieties have improved in this respect. Sow in a greenhouse in modules or blocks in late February. Slips or suckers can be bought in or pulled off established plants in the spring.

Planting and Growing: plants should be ready for planting out in late May at 900mm (36in) apart. Slips or suckers can be planted at 150–200mm (6–8in) deep and should be well firmed-in. The soil should be deeply worked and may benefit from sub-soiling, as the plant is deep rooted. Some inter-cropping can be possible in the first year. They are sensitive to frost and frost pockets should be avoided. They may need to be protected using a loose layer of straw, which is removed in the spring.

Soil Fertility and Rotation: a good-textured well-drained soil is essential for globe artichokes. The land should be well supplied with organic matter and manure or compost should be incorporated prior to planting, with an additional annual mulching in the spring.

Irrigation: artichokes require frequent irrigation during the growing season, as moisture deficit will result in poor quality heads. They will not, however, tolerate standing in water.

Weed Control: as for any perennial crop, it is important that the land is clean and free from perennial weeds prior to planting. Once established, the plants will be reasonably competitive with weeds. Inter-row cultivations can be carried out effectively and mulching is another possibility, using a woven ground cover, black plastic or compost.

Pests: globe artichokes are usually trouble free.

Diseases: care should be taken to ensure any planting material is free from disease.

Harvesting and Storage: the plants will come into production in the second season. Harvesting is possible from July to November. The main head is cut first and smaller heads from side shoots will be produced later. They are cut within 25–40mm (1–1.5in) of the stalk, at the immature stage, when young, tender and compact. When the buds begin to open and spread, the bracts become brownish, tough and stringy. They should be handled carefully to avoid bruising the bud leaves. They should be cold stored for immediate sale but can store for one or two weeks at 0–2°C and a relative humidity of 95 per cent if necessary.

Jerusalem Artichokes (O, *Helianthus tuberosus*)

This tuberous vegetable is, confusingly, unrelated to the globe artichoke (*see* above) but is, in fact, a member of the sunflower family. This is another vegetable of limited demand that can be useful for direct marketing in the winter season. Due to its tall habit, it can be grown as a windbreak for more tender crops. Although it is a perennial, it is best treated as an annual crop, removing all tubers from the ground at each harvest, as any remaining tubers can become weeds in following crops.

Seed: Jerusalem artichokes are grown from tubers. Smooth varieties are available, such as Fuseau, which are easier to prepare in the kitchen than the traditional knobbly types.

Planting and Growing: Jerusalems are easy to grow and can be planted as potatoes using the same equipment. They should be spaced 300mm (12in) apart in the row at any time from February to April, at a depth of 100–150mm (4–6in), in rows 750mm (30in) apart. The ridges can be earthed-up in the same way as potatoes. It is advisable not to let them grow too tall and to trim them to about 1.8m (6ft) high, so that energy can be put into tuber production. In the autumn, once leaves are starting to senesce, they can be cut down to just above ground-level and the tops removed for composting. This prevents gales from blowing them over and exposing the tubers.

Soil Fertility and Rotation: they will grow in most soils, but a well-drained deep loam is probably best. Manure as for potatoes.

Weed Control: if weeds are controlled early on, which should be by knocking down the ridges using harrows and re-ridging as for potatoes, then the crop will smother out weeds once they are established.

Pests: they suffer from few pests, though slugs may eat into the tubers, particularly in a wet, mild winter. Pheasants seem to find them a delicacy.

Diseases: few of note.

Harvesting and Storage: harvest can start in the autumn after the first frosts. Depending on scale of production this can either be by hand using a fork, or by using a potato harvester or spinner. Care is needed to avoid bruising or damaging the roots, as the skin is very thin. They are not damaged by frost and can be harvested as required from the field. They can keep for several months under cold storage with high humidity. This can be useful at the end of the spring when tubers will start to sprout if left in the field or to avoid trafficking on the field during harvest. Unless you have pigs, which love them, it is important to ensure all tubers or parts of tubers are removed, as volunteers can be a nuisance in subsequent crops.

Chapter 13

Sources of Information and Advice

Vegetable growing is a rewarding but challenging endeavour. All growers face a rapidly changing and evolving technical and business environment. Access to good quality and reasonably priced information is vital for successful vegetable production. All vegetable-production businesses depend on gathering information and advice to help them to decide what they want to achieve, how they want to manage their production enterprises and how to develop them further. Information and advice will be needed on a day-to-day basis (for example, specific information on a pest or disease) and over the longer term (such as for planning rotations or new marketing strategies). Studies have also shown that starting a vegetable-production business from scratch or converting from conventional to organic production is likely to involve a steep learning curve during the first three to five years whilst a new farming system is being developed and established. In this chapter we outline some of the more common methods of obtaining advice and information and provide a range of pointers to obtaining it.

INFORMATION CHANNELS

Although specific circumstances and needs vary and sources of information are open to rapid change over short periods of time, information and advice is, basically, available on a person-to-person basis, through the farming press (including radio and TV), through leaflets and technical guides and in books. Traditionally, much of this information has been perceived as being generated by researchers and as being transmitted to farmers and growers via journalists, through advisors, or at meetings, open days and agricultural shows.

However, it has become increasingly apparent that growers carry out a lot of research on their own farms in order to develop and stay in business.

Indeed, most growers find open days a valuable source of their information, not because scientists are present but because they are a good opportunity to meet and exchange ideas with other growers. Open days have become increasingly structured to allow growers to talk to each other and exchange information in this respect. Grower groups have also been formed and often meet at regular intervals to exchange information and views. Farming shows, although decreasing in popularity, are also a valuable form of interaction and information gathering, especially from commercial companies selling farming machinery and services.

Although spoken and printed materials are by far the most common method of receiving information, information is increasingly becoming available as audio-visual media, for instance as videos, CDs and DVDs. In this context the internet is also rapidly becoming the flexible media of choice for disseminating up-to-date technical and marketing information, as it is relatively cheap and potentially available to most farmers and growers. In this respect, all farm businesses should at least consider the benefits of having access to the internet and arranging for at least some form of basic training in its use.

SOURCES OF INFORMATION

Farming Press: the farming press is generally a good source of current information on a daily, weekly or monthly basis. Generally, press articles will give an overview of the technical information, which can be followed up in detail using other sources or by personal contact. It is also where adverts and trade classifieds are concentrated, both of which can be a valuable source of second-hand machinery or other goods.

The press aimed at farmers includes frequent radio programmes (for example, *Farming Today*), weekly newspapers (such as *Farmers Guardian*) or magazines (say *Grower* or *Farmers Weekly*), monthly magazines (such as *Vegetable Farmer*) or even quarterly magazines (for example *Organic Farming* or *Permaculture*). If you are unwilling to subscribe to some of these periodicals, it might be possible to ask a local library to take them.

Many research-based organizations, such as HDRA, Warwick HRI, HDC and EFRC, also produce occasional bulletins or other publications that summarize current research findings or other relevant materials. Much of this material can be freely obtained by registering with the organizations concerned. In some cases a small fee might be payable (for instance, NIAB associates scheme) but otherwise a lot of the material can be free. Many of these organizations also maintain websites, which they keep more or less up-to-date and which at least give an overview of new developments.

Open days or farm walks are a good source of information

Open Days, Shows and Seminars: most research organizations and agricultural colleges run open days for farmers as part of their obligation to disseminate research findings. Similarly, commercial companies (such as seed companies, machinery companies) also run open days to demonstrate their goods and services. Most of these are advertised in the farming press or available as press releases. Keep an eye on the farming press and other sources of information for the dates.

Most counties or regions hold dedicated agricultural shows on an annual basis. There are also a number of nationally established, annual shows aimed at vegetable growers (for example Vegex) and farmers in general (such as the Royal Show). These can be valuable as a source of information and for making contacts, especially regarding marketing opportunities and buying equipment.

Farmer Groups: most areas also have a number of established farmer groups. These can be structured around topics (soil fertility, biodynamics) or farm types (such as organic or dairy groups). These are often self-run, but can also be based around a local college or other institutions. EFRC supports an organic farm demonstration network and other research networks also exist. The Soil Association also supports a network of Organic Producer Groups around the country (*see* Useful Addresses) and other certification bodies often make provision to support their growers in various ways. All such groups can be a valuable way to make contacts and learn new information. Farmers' groups often also run open days and farm walks that enable farmers to visit other holdings in their area.

Advisory Services: advisory services generally offer person-to-person advice on aspects of organic farming. They normally involve the grower contracting the services of a specific advisor who visits the farm and gives advice to the farmer based on this. Advice may be varied and would generally cover all aspects of the farm business, from technical issues to help with obtaining relevant farming scheme payments. In addition to the visit, advisors can also help access other services, such as soil analysis or pest- and disease identification services, and advisors should be expected to keep themselves up-to-date with the latest farming policies, schemes and research. Farmers would normally expect to pay advisors sufficient to enable them to do this as well as the advisory work, although this cost would naturally be spread over all the farmers to which a particular advisor is providing advice. (For a list of organic advisory services *see* Useful Addresses.)

Many advisors belong to formal advisory groups or companies. These may also offer training workshops or provide other learning opportunities (such as open days or farm walks). One such group, the Organic Advisory Service (OAS) based at Elm Farm Research Centre, provides advice and information to commercial organic farmers. They offer on-farm advice and a telephone back-up service. They have a demonstration farm network and run a series of events and seminars, including a Horticultural Farmers Group. Other groups, such as ABACUS, also run a similar and active advisory service for organic farmers. The Organic Centre Wales (OCW) is the focal point for organic food and farming information in Wales, based at the University of Aberystwyth and run jointly by ADAS, EFRC, Institute of Grassland and Environmental Research (IGER), Institute of Rural Sciences and the Soil Association. Other organizations offer advice to both organic and conventional farmers.

Advice and information concerning conversion to organic farming can be obtained from the Organic Conversion Information Service (OCIS) help-line. Farmers and growers in any area of England can arrange for a free half-day visit and report, with a follow up full-day visit and expanded report, by an advisor experienced in organic production and marketing, who will provide impartial advice relevant to the business. The visits are currently provided by the OAS. Similar schemes operate in other parts of the UK on a country-by-country basis.

Certification Bodies: the various certifying bodies (*see* Useful Addresses) provide a certification service to organic farmers. Such a service ensures that

Many research institutes offer open days and seminars for farmers and growers

consumers are receiving produce from land that is farmed organically and helps form a bond of trust between farmers and consumers (*see* Chapter 1), but can also help farmers to gain further knowledge about their farming methods. Certification bodies generally charge for this service. However, in addition to this, many also provide some form of advice service. Such advice is usually provided over the telephone. It is especially useful for advice regarding interpretation of the organic standards and will be indispensable in situations where the standards are unclear on specific farming methods or techniques.

Research Institutes: a number of research institutes also exist in the UK. These institutes can either exclusively research organic farming (for instance, HDRA or EFRC) or can carry out research into both conventional and organic farming (for example, ADAS or Warwick HRI). Research is generally commissioned by the government, usually through DEFRA, but also by the Horticultural Development Council (HDC), which receives its money through a levy system on horticultural producers. Organic agricultural research covers a wide range of topics, from technical through to marketing and policy. Most research projects currently have at least some obligation to disseminate their findings to farmers and growers and many do this through the various channels already mentioned.

DEFRA: DEFRA is the government department ultimately responsible for policy and research on organic agriculture in the UK. Regional development agencies and devolved assemblies also have some power to set agricultural policies and guide development in the regions. All these institutes provide information in various forms and various ways. The most accessible and up-to-date way of obtaining information from these bodies is to access their websites, where it is provided free. Many also have local offices (*see* Useful Addresses).

Leaflets and Technical Guides: many of the institutes and organizations mentioned above also produce a range of leaflets and factsheets aimed at farmers and growers. These generally aim to provide practical information to farmers and growers and cover all aspects of farming. They include technical guides, marketing guides and policy guides, among others. The websites of many of the institutes will provide free access to some of this material, some is available free of charge by post and some is available for a (usually small) fee.

Commercial companies also often produce this type of material, especially seed and machinery companies. Some of the catalogues produced by these companies also contain a good deal of technical information and even the results of recent research carried out by the company or research institutes. Once again, much of this material will be free at agricultural shows, by registering an interest with the company, or from their websites.

USING INFORMATION

Collecting information is normally fairly straightforward and, in fact, there is currently a huge amount of information generated between all the various sources listed above. Much of this will often be contradictory and, anyway, not strictly applicable in specific farm circumstances. This can create a feeling of information overload and it is not easy to provide an answer to this.

Experience indicates that the best way to keep track of information is to be systematic. Here we can only provide a few pointers as to how to do this but, in the end, each person will have to develop a system that suits them.

- Keep notes and farm records. It is vitally important to keep good farm notes and records so that any unexpected outcomes can be explained. Often, fortuitous accidents can lead to better ways of doing things and, with good records, these opportunities can be developed. It is also important to keep records in order to be able to get an understanding of how your farm system is developing, for instance whether a weed problem is getting worse or improving over time and through a rotation. Modern technology, such as computers and digital cameras, can significantly aid the process of keeping records. One picture may save a page of notes.

- Find a few favourite sources of information that suit you and that provide information in a form that you want it. This may, for instance, be a farming magazine, an advisor or website. Keep up to date with it but don't worry if some of it passes you by in busy periods. Other sources of information include research projects and technical bulletins, but may also include commercial companies. Try to match some of this information with your farm records and notes and use it to develop new technical skills or to provide solutions to existing problems.
- Attend some open days and seminars, especially local ones. Many farmers and growers find such events to be a good means of visiting other farms and sharing information.
- Join a farmers' group. Many farmers and growers report that they have positive experiences with farmers' groups, as they are a forum for sharing information and problems and can be a source of technical innovation.
- Take advantage of government training schemes to acquire new skills or consolidate information. Many local and agricultural colleges offer training to the land-based sector and many government grants can be used to, at least partially, cover the costs of these courses. Such training is a good way to consolidate experience and put it in some context even if you 'already know it'.
- Take time occasionally to reflect on the business and put all this information in context. This can be done either individually, with an advisor or with a group. Often solutions to problems present themselves when a person steps back from the immediate business in hand and reflects on the history or development of a particular situation.

Chapter 14
Future Prospects

Vegetable production has changed enormously over the past century. Think of the kitchen gardens of the nineteenth century or the market gardens around towns in the early twentieth century, supplying wholesale markets or market towns, as compared to the modern, large and world-wide distribution networks on which supermarkets currently depend. Much of the change to this large-scale production has come about in the last few decades, which is also the time in which organic vegetable production has come to the forefront. To some extent, changes in organic horticulture have mirrored those in conventional production, changing from the dominance of small-scale production in the south-west of England and Wales to include large-scale producers in the specialist vegetable growing areas of the country.

However, organic production occupies a unique position within agriculture, in that it offers a vision of what we should be doing to guarantee the sustainability of not only our production systems but also our farming communities. In a sense, organic farmers and consumers have taken a stand and have defined, at least, the environmental limits beyond which it is not sensible to stray, if we are to ensure that future generations will also be able to produce food and support themselves. They are also beginning to build a consensus on the types of food-production systems that develop and satisfy social capital. That is, to allow people to develop equitable relationships in which both growers and consumers have space to develop their aspirations.

If one thing is clear, it is that horticulture will continue to change. Although it is always difficult to predict what will happen in the future, we have provided a brief insight into where we think organic horticulture will go in the short- to medium-term.

Short Term

In the short term, organic horticulture is likely to continue along the same path, as demand for organic vegetables continues to grow. Supermarkets are likely to continue to dominate this trade for some years to come, as this style of shopping is currently in tune with modern urban lifestyles and, in any case, they are economically well-positioned to maintain control of supply chains. However, direct marketing through box schemes, farm shops and farmers' markets are likely to have an increasingly important role to play in offering the consumer an alternative. That is, we are likely to see small increases in direct sales year-on-year to meet the growing demand for local fruit and vegetable production. In the direct-sales market some schemes are likely to continue growing and some of them are now very large schemes indeed – in a sense, competing with supermarkets for the fresh-vegetable trade. This is not necessarily a direct threat to small farm based and local schemes, as they can still provide strong and distinctive products, with a farm story and direct connection between the consumer and grower.

All these trends will be influenced to a greater or lesser extent by government policies on the environment and health. They are also likely to be sensitive to food scares and media campaigns on issues such as school dinners or obesity. Health will become increasingly important in the short- to medium-term, as the UK population ages, the 'processed food generation' of today enters middle age and the consequences of their poor eating-habits are felt through increased health costs. Healthy eating is already beginning to become a focus of government policy initiatives. Organic growers are well placed to take advantage of schemes like the '5-a-day' initiative aimed at getting more people to eat five portions of vegetables or fruit a day. Schemes targeted at schools, such as the free fruit for schools initiative in primary schools, are likely to be extended to include fresh vegetables, as government tries to convince people to adopt healthy eating habits. Public procurement of local fresh produce is likely to increase in the short- to medium-term, as local councils strive to implement government policy. Health Trusts are also likely to begin to look to sourcing local and fresh produce.

One overlooked area is that of leisure. A growing proportion of disposable income is being spent on eating out and one challenge for organic growers is to make organic and nutritious food available to the public through outlets such as restaurants or other eating places. People are also increasingly looking for these options, as part of the experience of visiting tourist attractions or as part of other activities, such as county shows or music festivals, and demand for speciality produce, like heritage varieties, for the catering trade is likely to increase. Niche marketing of speciality foods is also likely to be popular as part of these cultural experiences.

In the short term, price is likely to be a major issue for organic producers and consumers. There is sufficient evidence to show that costs of organic vegetable production are higher than conventional production for a range of reasons, for example the need for fertility building during the rotation and higher seed and labour costs. There is, therefore, a perception on the part of consumers that, although they want to buy organic food, the price is too high. This has led to a downward pressure on price, as supermarkets, in particular, try to satisfy the demand. This pressure is also a result of increased competition within the organic vegetable market. Low prices, in turn, affect the long-term viability of organic vegetable businesses. The challenge for organic growers is therefore to make organic food accessible to all at a price that allows them to make a living. This can be done by not only reducing unnecessary costs, but also by convincingly justifying the cost of organic produce by exposing the hidden costs of conventional food and stressing the environmental benefits of organic food and the health benefits of fresh, local food. These arguments are more likely to be heard if producers cooperate with each other and form alliances with consumer interest groups.

In the short term, government environmental policy is likely to be largely driven by reform of the EU Common Agricultural Policy, which is beginning to switch payments from subsidies for production to subsidies for providing environmental services. Whilst vegetable growers have never benefited hugely from this system, they are likely to be able to get increasing access to some of these payments. Government policy, delivered through DEFRA, is also looking to deliver a rural landscape that urban populations expect. Organic growers are likely to be able to benefit from programmes aimed at promoting the rural economy, as research begins to show the valuable role that organic vegetable-production can play in maintaining rural jobs and services.

All these initiatives will bring increased opportunities for organic vegetable production, both for existing growers and new businesses. In the face of these complex changes, growers will need increased access to information in order to run successful businesses. This will include market information but also technical information, pest and disease forecasts, and financial and advisory services.

Medium Term

Biodiversity and environment will become an increasing priority for both growers and governments in the medium term. The backdrop is likely to be increasing prices of oil and other scarce resources, including water, coupled to an increasingly impoverished flora and fauna. On the positive side, these pressures are likely to reinforce the trend towards organic produce and local food-production systems, as the environmental advantages of short food chains begin to become practical and necessary. Local authorities and businesses are likely to look to local sourcing, as the costs of transport rise. However, there will also be commercial pressure for GM and labour-saving technologies to be adopted by farmers in order to keep the costs of food down. Any investment or adoption of GM technology by conventional farmers could have negative consequences for organic growers, both in terms of diverting research and development resources away from organically useful technologies and the unresolved question marks over co-existence.

All this is likely to be against a background of increasing climate uncertainty. Weather events in the UK and northern Europe are likely to become more extreme. For instance, more hail damage can be expected in the summer and longer periods of drought or excessive rainfall will become more common. This will obviously have an impact on vegetable-production practices. There will also be consequent changes in weed, pest and disease flora and fauna. Warmer winters will allow continual germination and growth of weeds and is likely to allow pests and diseases to bridge from one season to another – or even allow new ones to become established. In the longer-term, climate changes may even alter the types of vegetable crops that can be sensibly grown. For growers, the basic challenge will be to design biodiverse systems that are resilient in the face of environmental and economic shock.

Long Term

In the long term, all we can say is that there will be a need for fresh, high quality produce in whatever future scenarios unfold. We also confidently assert that the organic approach to production of vegetables will place organic growers at the forefront of food production in any of these future scenarios. The principles of organic agriculture, and of organic vegetable production, as outlined in this book, will therefore be with us for some time to come, although the actual practices may change and develop in their detail. In any analysis, a system of agriculture that puts self-sufficiency, sustainability and human and animal welfare at its heart will be well placed to adapt to any change, be it social or environmental.

Further Reading

PERIODICALS

A range of periodicals aimed at farmers and growers is available. Also, there are farming programmes on local radio, as well as national radio and television. Five popular periodicals are;

The Grower (weekly, Tel: 01353 654422)
Farmers Guardian (weekly, Tel: 01858 43883)
Farmers Weekly (weekly, Tel: 0845 0777744)
Vegetable Farmer (monthly, Tel: 01622 695656)
Organic Farming (quarterly, Tel: 0117 929 0661)

In addition, certification bodies, research institutes, levy bodies and DEFRA publish newsletters or occasional bulletins on a wide range of farming issues (*see* Useful Addresses).

LEAFLETS AND TECHNICAL GUIDES

A large amount of technical information is published through the periodicals (*see* above). In addition, commercial companies often publish guides to their products that often contain a good deal of information, especially seed and machinery catalogues. Contact the companies individually. Research leaflets are available through research institutions and usually published on an ad hoc basis (*see* Useful Addresses). Some useful and currently available leaflets are given below:

'A guide to better soil structure', Guideline from the National Soil Resources Institute (Cranfield University, Silsoe, 2002), 19pp.
'A guide to managing crop establishment', Guideline from the UK Soil Management Initiative (SMI), 51pp.
Azeez, G., 'The biodiversity benefits of organic farming', Report (Soil Association, 2000)
Edwards, M.A., 'Using green waste compost in agriculture', (HDRA, 1999), 19pp.

'Farm Gate Sales to the Public', NFU Legal Guide, 2nd edition (Shaw & Sons)

Gladders, P., Davies, G., Wolfe, M., Haward, R., 'Diseases of organic vegetables' (ADAS: Wolverhampton, 1992), 78pp.

Pilley, G., 'A share in the harvest – a feasibility study for community supported agriculture' (Soil Association, 2004), 43pp.

Shepard, M. and Gibbs, P. with Philips L., 'Managing manure on organic farms'. DEFRA booklet (ADAS and EFRC, 2002), 23pp.

'Soil Management on Organic Farms', A guide from the Soil Association (SA, 2003), 39pp.

'The UK Organic Vegetable Market', Report (HDRA, 2004).

Soil Association technical guides are available from the Soil Association Producer Services on a range of subjects including:

'Organic Carrot Production'
'Organic Onion Production'
'Organic Potato Production'
'Rotations for Organic Horticultural Field Crops'
'Setting Up an Organic Box Scheme'
'Soil Management on Organic Farms'.

Soil Association factsheets are available free of charge to producer members and downloadable from www.soilassociation.org/ps
Titles include:

'Organic Horticulture: an introductory guide' (2003)
'Composts, plant raising media, soil inoculants and mulches for use in organic production' (2004)
'Marketing Information for Organic Growers (Horticultural Crops)' (2004)
'Fertilisers for use in organic production' (2004)
'Slug Control in Organic Systems' (2004)
'Materials for pest and disease control in organic crops' (2004)
'Organic Herbs' (2004)
'Organic ornamental plants' (2004)
'Growing organic vegetables for a box scheme' (2004)
'Suppliers of transplants, herbs and root stock for use in organic systems' (2004)

Information on vegetable variety-performance is constantly being updated as new varieties are produced and marketed. Apart from seed catalogues, a range of more impartial sources include:

- NIAB produce the Organic Vegetable Handbook, Vegetable Variety Handbook and Potato Variety Handbook. Contact NIAB (www.niab.com)

- Varietal trial factsheets. Available from HDC (www.hdc.org.uk) or NIAB (www.niab.com)
- Soil Association Producer Guides (*see* above or www.soilassociation.org)
- Organic seed availability at the Centre for Organic Seed Information (www.cosi.org.uk) or OrganicXseeds (www.organicXseeds.com)

BOOKS

A valuable source of technical information and usually a source of more detailed background information. Some informative texts are given below:

Blake, F., *Organic Farming and Growing* (Crowood, 1994), 221pp.
Coleman, E., *The New Organic Grower* (Cassell, London, 1989), 269pp.
Lampkin, N., *Organic Farming* (The Farming Press, Ipswich, 1990)
Lampkin, N., Measures, M., Padel, S., *The Organic Farm Management Handbook* 6th edition (University of Wales, Aberystwyth, 2004)
Newton, J., *Profitable Organic Farming* (Blackwell, Oxford, 1995), 142pp.

INTERNET

Rapidly changing forum for exchange of information and views. Users should search the internet using one of the numerous search engines that are available and earmark sites that are useful to them. Some sites provide (at least some) free up-to-date information and are given below:

ADAS website deals with organic issues, including water quality and plastic disposal – www.adas.co.uk/home/organic.html
Centre for Organic Seed Information – www.cosi.org.uk
DEFRA, Organic Food and Farming – www.defra.gov.uk/farm/organic/default.htm
Elm Farm Research Centre – www.efrc.com
Environmentally sensitive farming – www.environmentsensitivefarming.co.uk
HDRA Organic Vegetable Systems Studies – www.organicveg.org.uk
HDRA Organic Weed Management – www.organicweeds.org.uk
Horticultural Development Council (HDC) – www.hdc.org.uk
National Sustainable Agricultural Information Service (US site) – www.attra.org
NIAB – www.niab.com
Organic Centre Wales – www.organic.aber.ac.uk/index.shtml
Organic Research (CABI) – www.organic-research.com

Organic-X-seeds – www.organicxseeds.co.uk
Scottish Agricultural College – www.sac.ac.uk
Soil Management Initiative – www.smi.org.uk
The Organic Marketplace – www.soilassociation.org/organicmarketplace
Warwick HRI – www2.hri.ac.uk

SUPPLIERS

It is not possible, nor appropriate, to give an exhaustive list of horticultural suppliers in a book such as this. All growers will naturally develop a list of suppliers with whom they can work, as they develop their business. Many of the periodicals mentioned above have a large number of advertisers and some produce summaries from time to time. A useful source of information on suppliers of horticultural sundries is the horticultural supplier guide issued with the *Grower* magazine at Christmas. The Commercial Horticultural Association has an online buyers guide (www.cha-hort.com/Buyers-guide). Many of the fact sheets supplied by the Soil Association also list suppliers (*see* above).

Useful Addresses

ADVICE – CONVERSION

The Organic Conversion Information Service (OCIS)
Advice and information concerning organic farming can be obtained from the Organic Conversion Information Service (OCIS) helpline. Farmers and growers in any area of England can arrange for a free half-day visit and report, with a follow up full-day visit and expanded report, by an advisor experienced in organic production and marketing who will provide impartial advice relevant to the business. The visits are provided by the Organic Advisory Service (OAS), who are based at Elm Farm Research Centre (EFRC).
OCIS Helpline
(England) – 0117 922 7707

Organic Centre Wales (OCW)
The OCW delivers OCIS in Wales, supported by ADAS and the OAS. An information pack and up to two free advisory visits are provided, plus detailed conversion-planning linked to the Farming Connect Farm Business Development Plans.
Organic Centre Wales Technical Helpline – 01970 622100
Email: organic-helpline@aber.ac.uk

Scottish Agricultural College
Free telephone advice and information to converting and existing organic farmers from a network of local advisory offices. Funding is available towards the cost of advisory help in preparing conversion plans. Face-to-face consultations are available on a charged basis.
Scottish Agricultural College
Organic Helpline – 01224 711072

Northern Ireland – Department of Agriculture and Rural Development (DARD)
DARD has an advisory team based at Greenmount College and offers OCIS.
Field Scale Vegetables
Helpline – 028 9442 6683
Market Gardening – 028 9442 6765

ADVICE – ORGANIC GENERAL

Abacus Organic Associates
61 Raby Park Road
Neston, South Wirral
Cheshire CH64 9SW
Tel/Fax: 01780 721019
Email:
stephen.briggs@abacusorganic.co.uk
Website: www.abacusorganic.co.uk

ADAS National Help Desk
Tel: 0845 766 0085
Email: enquiries@adas.co.uk
Website: www.adas.co.uk

Greenmount College of Agriculture and Horticulture
Antrim
Northern Ireland BT41 4UP
Tel: 02894 426601 or
Freephone 0800 0284291
Fax: 02894 426606
Email: enquiries@dardni.gov.uk
Website: www.greenmount.ac.uk

The Organic Advisory Service (OAS)
The OAS is based at Elm Farm Research Centre and, in addition to OCIS, provides advice and information to commercial organic farmers. They offer on-farm advice and a telephone back-up service. They have a demonstration farm network and run a series of events and seminars. There is a Horticultural Farmers Group.

The Organic Advisory Service (OAS)
Elm Farm Research Centre
Hamstead Marshall
Newbury
Berkshire RG20 0HR
Tel: 01488 658298
Fax: 01488 658503
Email: oas@efrc.com
Website: www.efrc.com

The Organic Centre (Ireland)
Rossinver
Co. Leitrim
Republic of Ireland
Tel: 00 353 71 98 54338
Fax: 00 353 71 98 54343
Email: organiccentre@eircom.net
Website: www.theorganiccentre.ie

Organic Centre Wales (OCW)
The OCW is the focal point for organic food and farming information in Wales, based at the University of Aberystwyth and run jointly by ADAS, EFRC, Institute of Grassland and Environmental Research (IGER), Institute of Rural Sciences and the Soil Association.

Organic Centre Wales (OCW)
University of Wales
Aberystwyth
Ceredigion SY23 3AL
Tel: 01970 622248
Fax: 01970 622238
Email: organic@aber.ac.uk
Website: www.organic.aber.ac.uk

Scottish Agricultural College (SAC) Organic Farming Unit
Ferguson Building
Craibstone Estate
Aberdeen AB21 9YA
Tel: 01224 711072
Fax: 01224 711293
Email: david.younie@sac.co.uk
Website: www.sac.ac.uk/consul tancy/services/organicservices/

Soil Association Producer Services
Bristol House
40–56 Victoria Street
Bristol BS1 6BY
Tel: 0117 914 2400
Fax: 0117 925 2504
Email: ps@soilassociation.org
Website:
www.soilassociation.org/ps

There are a number of organic centres linked to the Soil Association including:

Organic South West
Tel: 01579 371147
Email: osw@soilassociation.org
Website:
www.organicsouthwest.org

Northwest Organic Centre
Tel: 01995 642206
Email:
enquiries@nworganiccentre.org
Website:
www.nworganiccentre.org

Soil Association Scotland
Tel: 0131 666 2474
Email: contact@sascotland.org
Website:
www.soilassociationscotland.org

Yorkshire Organic Centre
Tel: 01756 796222
Email:
info@yorkshireorganiccentre.org
Website:
www.yorkshireorganiccentre.org

North East Organic Programme
Tel: 0845 121 7645
Email: neop@northeastorganic.org
Website:
www.northeastorganic.org

CERTIFICATION BODIES

Organic Farmers and Growers Ltd
The Elim Centre
Lancaster Road
Shrewsbury
Shropshire SY1 3LE
Tel: 01743 440512
Fax: 01743 461441
Email: info@organicfarmers.uk.com
Website:
www.organicfarmers.uk.com

Scottish Organic Producers Association
Scottish Organic Centre
10th Avenue
Royal Highland Centre
Ingliston
Edinburgh EH28 8NF
Support & Development:
Tel: 0131 333 0940
Fax: 0131 333 2290
Certification:
Tel: 0131 335 6606
Fax: 0131 335 6607
Email: sopa@sfqc.co.uk
Website: www.sopa.org.uk

Organic Food Federation
31 Turbine Way
Eco Tech Business Park
Swaffham
Norfolk PE37 7XD
Tel: 01760 720444
Fax: 01760 720790
Email: info@orgfoodfed.com
Website: www.orgfoodfed.com

Soil Association Certification Ltd
Bristol House
40–56 Victoria Street
Bristol BS1 6BY
Farmers and Growers:
Tel: 0117 914 2406
Processors:
Tel: 0117 914 2407
Fax: 0117 925 2504
Email:
prod.cert@soilassociation.org
Website: www.soilassociation.org

Bio-Dynamic Agricultural Association
The Painswick Inn Project
Gloucester Street
Stroud GL5 1QG
Tel: 01453 759501
Fax: 01453 759501
Email:
bdaa@biodynamic.freeserve.co.uk
Website:
www.biodynamic.org.uk

Irish Organic Farmers and Growers Association
Harbour Building
Harbour Road
Kilbeggan
Co. Westmeath
Republic of Ireland
Tel: 00 353 506 32563
Fax: 00 353 506 32063
Email: iofga@eircom.net
Website: www.irishorganic.ie

Organic Trust Limited
Vernon House
2 Vernon Avenue
Clontarf
Dublin 3
Republic of Ireland
Tel: 00 353 185 30271
Fax: 00 353 185 30271
Email: organic@iol.ie
Website: www.organic-trust.org

CMi Certification
Long Hanborough
Oxford OX29 8LH
Tel: 01993 885651
Fax: 01993 885611
Email:
enquiries@cmicertification.com
Website: www.cmi-plc.com

Quality Welsh Food Certification Ltd
Gorseland
North Road
Aberystwyth
Ceredigion SY23 2WB
Tel: 01970 636688
Fax: 01970 624049
Email: mossj@wfsagri.net

Ascisco Ltd
Bristol House
40–56 Victoria Street
Bristol BS1 6BY
Farmers and growers:
Tel: 0117 914 2406
Processors:
Tel: 0117 914 2407
Fax: 0117 925 2504
Email:
DPeace@soilassociation.org

RESEARCH/LEVY BODIES

ADAS Terrington
Terrington St. Clement
King's Lynn
Norfolk PE34 4PW
Tel: 01553 828621
Fax: 01553 827229
Email: bill.cormack@adas.co.uk
Website:
www.stocklessorganic.co.uk

British Potato Council (BPC)
4300 Nash Court
John Smith Drive
Oxford Business Park
Oxford OX4 2RT
Tel: 01865 714455
Fax: 01865 782231
Website: www.potato.org.uk

Central Science Laboratory (CSL)
Sand Hutton
York YO41 1LZ
Tel: 01904 462000
Fax: 01904 462111
Email: science@csl.gov.uk
Website: www.csl.gov.uk

Elm Farm Research Centre (EFRC)
Hamstead Marshall
Newbury
Berkshire RG20 0HR
Tel: 01488 658298
Fax: 01488 658503
Email: elmfarm@efrc.com
Website: www.efrc.com

HDC – Horticultural Development Council
Bradbourne House
East Malling
Kent ME19 6DZ
Tel: 01732 848383
Fax: 01732 848498
Email: hdc@hdc.org.uk
Website: www.hdc.org.uk

HDRA – The Organic Organisation
Research and Development Division
Ryton Organic Gardens
Coventry CV8 3LG
Tel: 024 7630 8200
Fax: 024 7663 9229
Email: research@hdra.org.uk
Website:
www.hdra.org.uk/research

Institute of Rural Sciences
Organic Farming Research Unit
University of Wales
Aberystwyth
Ceridigion SY23 3AL
Tel: 01970 622248
Fax: 01970 622238
Email: organic@aber.ac.uk
Website: www.irs.aber.ac.uk

Nafferton Ecological Farming Group
Nafferton Farm
Stocksfield
Northumberland NE43 7XD
Tel: 01661 830222
Fax: 01661 831006
Email: tcoa@ncl.ac.uk
Website: www.ncl.ac.uk/tcoa

National Institute of Agricultural Botany
Huntingdon Road
Cambridge
Cambs CB3 0LE
Tel: 01223 342200
Fax: 01223 277602
Email: info@niab.com
Website: www.niab.com

Organic Studies Centre
Duchy College
Rosewarne
Camborne
Cornwall TR14 0AB
Tel: 01209 722155
Fax: 01209 722156
Website:
www.organicstudiescornwall.co.uk

Warwick HRI
Wellesbourne
Warwick CV35 9EF
Tel: 02476 574455
Fax: 02476 574500
Website: www.warwickhri.ac.uk

MARKETING

BigBarn
Bigbarn Ltd
College Farm
Great Barford
Bedfordshire MK44 3JJ
Tel: 01234 871005
Email: ant@bigbarn.co.uk
Website: www.bigbarn.co.uk

F3 – the local food consultants
PO Box 1234
Bristol BS99 2PG
Tel: 0845 458 9525
Email: mail@localfood.org.uk

Farmers' Markets
Website: www.farmersmarkets.net

Food Standards Agency
Aviation House
125 Kingsway
London WC2B 6NH
Tel: 020 7276 8000
Website: www.food.gov.uk

Institute of Grocery Distribution
IGD, Grange Lane
Letchmore Heath
Watford
Hertfordshire WD25 8GD
Tel: 01923 857141
Email: igd@igd.com
Website: www.igd.com

National Farmers' Retail and Markets Association (FARMA)
PO BOX 575
Southampton SO15 7BZ
Tel: 0845 230 2150
Website: www.farmersmarkets.net/ or www.farmshopping.com

NFU
Agriculture House
164 Shaftesbury Avenue
London WC2H 8HL
Tel: 020 7331 7200
Email: nfu@nfuonline.com
Website: www.nfu.org.uk

Organic Monitor
79 Western Road
London W5 6DT
Tel: 020 8567 0788
Website:
www.organicmonitor.com

Organic TS
Tel: 07974 103109
Email: info@organicTS.com
Website: www.organicts.com

Realproduce
Online food ordering from local producers.
Website: www.realproduce.co.uk

Scottish Agricultural College
Ferguson Building
Craibstone Estate
Aberdeen AB21 9YA
Tel: 01224 711072
Fax: 01224 711293
Website: www.sac.ac.uk/management/external/diversification/Valueadd/farmshop.asp

Scottish Association of Farmers' Markets
Website:
www.scottishfarmersmarkets.co.uk

Sustain: the alliance for better food and farming
CPAG Building
Second Floor

White Lion Street
London
Tel: 020 7837 1228
Website: www.sustainweb.org

OTHER ORGANIZATIONS

European Weed Research Society
Physical and cultural weed control.
Website: www.ewrs.org/pwc

The Waste and Resources Action Programme – Organics
The Old Academy
21 Horse Fair
Banbury
Oxon OX16 0AH
WRAP Helpline: Freephone 0808 100 2040
Email: helpline@wrap.org.uk
Website: www.wrap.org.uk/materials/organics

World Wide Opportunities on Organic Farms (WWOOF)
PO Box 2675
Lewes
East Sussex BN7 1RB
Tel/Fax: 01273 476 286
Email: hello@wwoof.org
Website: www.wwoof.org.uk

EDUCATION/TRAINING

Many local and national colleges and universities run education and training workshops in aspects of farming and land management. They should be accessible through your local government or information services. Others include:

Abacus Organic Associates
(*see* above)

Centre for Alternative Technology
Macchynlleth
Powys SY20 9AZ
Tel: 01654 705981
Fax: 01654 703605
Email: info@cat.org.uk
Website: www.cat.org.uk

Elm Farm Research Centre
(*see* above)

IOTA – Institute of Organic Training and Advice
IOTA is a professional body for trainers, advisors and other extension workers involved in organic food and farming.
Cow Hall
Newcastle
Craven Arms
Shropshire SY7 8PG
Tel: 01547 510344
Email:
diana@iota.wanadoo.co.uk
Website: www.aber.ac.uk/iota

LANTRA – the Sector Skills Council for the Environmental and Land-Based Sector
Lantra House
Stoneleigh Park, Nr Coventry
Warwickshire CV8 2LG
Tel: 024 7669 6996
Fax: 024 7669 6732
Email: connect@lantra.co.uk
Website: www.lantra.co.uk

Organic Centre Ireland
(*see* above)

Organic Centre Wales (*see* above)

GOVERNMENT/GRANTS

DEFRA facilitates access to grants, training and education for the farming sector. In addition to the main centre below there are numerous regional development offices that can also be contacted for information.

DEFRA
Information Resource Centre
Lower Ground Floor
Ergon House, c/o Nobel House
17 Smith Square
London SW1P 3JR
Defra helpline: 08459 33 55 77
Email: helpline@defra.gsi.gov.uk
Website: www.defra.gov.uk

Defra Rural Development Service for Organic Entry Level Scheme and Higher Level Scheme including East of England, East Midlands, North West, South East, South West, West Midlands, Yorkshire and Humber.

English Nature
Northminster House
Peterborough PE1 1UA
Tel: 01733 455101
Email:
enquiries@english-nature.org.uk
Website: www.english-nature.org.uk

Farming Connect Wales
Tel: 08456 000813
Website: www.wales.gov.uk/farm ingconnect

Rural Payments Agency
RPA Customer Service Centre
Tel: 0845 603 7777
Email: customer.service.centre@ rpa.gsi.gov.uk
Website: www.rpa.gov.uk

Scottish Executive Environment and Rural Affairs Department
Pentland House
47 Robb's Loan
Edinburgh EH14 1TY
Tel: 0131 556 8400
Fax: 0131 244 6116
Email: ceu@scotland.gsi.gov.uk
Website: www.scotland.gov.uk

Welsh Assembly Government Department for Environment, Planning and the Countryside
Crown Building
Cathays Park
Cardiff CF10 3NQ
Tel: 02920 825111
Email:
agriculture@wales.gsi.gov.uk
Website:
www.countryside.wales.gov.uk

Index